the safety training NINJA

the safety training NINJA

Regina McMichael

American Society of Safety Professionals, 520 N. Northwest Highway, Park Ridge, IL 60068

Print ISBN: 978-0-939874-25-5
E-book ISBN: 978-0-939874-26-2

Managing Editor: Rick Blanchette, ASSP
Text design and composition: Cathy Lombardi
Cover design: Amy Ackerman, ASSP

Printed in the United States of America

27 26 25 24 23 22 21 20 19 1 2 3 4 5 6 7 8

Library of Congress Cataloging-in-Publication Data

Names: McMichael, Regina, (Safety Speaker), author.
Title: The safety training ninja / Regina McMichael.
Description: Park Ridge, IL : American Society of Safety Professionals, 2019.
Identifiers: LCCN 2018043933| ISBN 9780939874255 (paperback : alk. paper) | ISBN 9780939874262 (ebook)
Subjects: LCSH: Safety education, Industrial. | Industrial safety.
Classification: LCC T55.2 .M395 2018 | DDC 658.3/1244--dc23 LC record available at https://lccn.loc.gov/2018043933

Contents

Acknowledgments

Special thanks go out to the following people:

Mike McMichael, who supported me building my business as I became the ninja I was meant to be.

Cooper McMichael, who used the word ninja for cool stuff when he was younger and inspires me every day.

Geraldine Cooper, for unconditional love and support to first become a safety professional and then to continue to grow and find my voice.

Pandora Burge, for being kind to people and teaching me the power of cookies, recognition, and grace.

Kelley Edmier, for being the best training partner ever and the most natural training ninja there is.

Taylor Conlon, Kelsey Crawford, and Christina Leader for helping me sustain my business while I wrote this book and for editing all the heinous typos.

Anna Rice, for understanding my need for whimsy when creating the art for this book.

Abby Ferri, for her support, review, and teaching me stuff every time I talk to her.

Dan Snyder, for sharing a lot about ISD and supporting the ninja way.

Rich Gallagher, for being the first SME I converted to the ADDIE cycle and for giving me the confidence to convert more.

Tom Fogarty, for being a great member of my team during my formative years in ISD, and for teaching me a lot of stuff I didn't know, and for challenging me to be great.

Elaine Biech, who is the best book mentor and my first trainer on real ISD.

James Pomeroy, who asked me the important question about what I really wanted to do with my career and then supported my adventure!

Jim York, for being a great boss as I pursued my knowledge for ISD despite being a safety person and for letting me experiment with e-learning on his watch.

Pat Melzer and Sarah Londono, who taught me conference management and how to make all my learning stuff fit in those parameters.

Pete Chaney and Mike O'Brien, who let me first write and then teach with passion.

Anyone who ever attended ninja training, because you inspire me to be better with every question you ask and with the nodding affirmation you provide.

Introduction
How the Safety Training Ninja Was Born

My first training class was on how to back up state-owned vehicles safely. I was given a carousel of slides two days after I saw it presented for the first time. It was a mess, but I loved it. At university, if I was given the option to write a paper or give a speech, the choice was simple and immediate: speak! My mother likes to call it the gift of gab. This is a phrase reserved for use by people over the age of eighty, but she is right—communicating is my passion! I was born the last of six children in a military family, and I still hear the echoes of my father telling us—in the middle of a story—to get to the point. For all training, but particularly safety training, it is critical for our message to be clear, concise, and to the point. I hope you learn how to achieve that here.

Over the years, I self-learned many of the best learning development, design, and delivery strategies through trial and error. It wasn't until I took my first instructional design course that I found out what I did when I trained was part of a bigger vocation and officially a systematic approach to training. It was a relief to know I could take courses, read books, and learn from others. At that point, I joined the American Society of Training and Development (ASTD), now called the Association for Talent Development (ATD). I attend their conferences and learn something from every speaker I hear. I continue to attend their programs and use their education resources to improve my professional capabilities.

Ultimately, the ninja concept was born out of the love for the learning industry's use of creative titles like guru, evangelist, and champion. None of those felt right for me, and then I thought of my son, who was about twelve at the time. Many things that were impressive and cool to him got the title of ninja. I liked the sound of it, and, more importantly, it added some whimsy and creativity to a serious topic. I taught my first Safety Training Ninja class in 2014 for several different audiences. It was clear then that EHS professionals were hungry for a simple, effective approach to design, development, and delivery of better safety training materials.

Navigating This Book

I intentionally teach in an informal manner because I believe it helps learners connect to the materials and each other. I have written this book to match that style. If you have ever seen me teach you will likely hear my voice as your read. The text and materials are conversational. The material in the book will be new to many readers, and I wanted to make sure it was easy to understand and implement all or even just parts of the content. I share stories, tips, and tricks in every chapter, as well as offer some job aids and checklists to help you become the best safety training ninja you can be.

Training versus education—does it really matter what we call it? The learning industry has plenty of smart people to explain the nuanced differences between the two terms, but for EHS professionals, we tend to call what we deliver *training*. You may be developing actual education programs or both, but for the sake of simplicity, we call it training in this book.

This book assumes you are the trainer. You may be developing materials for others to deliver, and all the information in this book will yield a better program for them; however, developing programs for others to teach requires even more preparation, supporting materials, and content development. Perhaps that is my next book?

We call the people you are creating your training program for *stakeholders*. There are a lot of other names we could use, such as client, manager, or even owner, but since this book is designed for all EHS professionals, I stick with an encompassing term that can be accepted by all.

To avoid some of the inflammatory discussions about the differences between employees and workers and who is responsible for their training and education, I use the term *learner*. This also allows the discussion to include the many other terms used in the industry. I specifically avoid the term *student*. In almost all examples, we talk about training and educating adults, and since some adults might be discouraged by the term used for children, we adopted a learning industry standard for clarity, simplicity, and respect. There are many other terms that you can use in your programs, including *delegate* (popular in Europe and Asia), *attendee*, as well as any company or other stakeholder term that fits you best.

As you read the materials presented, you may wonder, "Can I use the system presented for all types of EHS training?" The answer is yes, but of course you may need to adjust the emphasis placed on the different phases of your effort. A toolbox talk may be shorter than a compliance-based course but that doesn't mean you should not be sure it does what its meant to do: help your learners learn what they need to know to do their jobs safely.

I don't discuss generational issues in EHS training. A well-built course will work for everyone if you utilize the concepts I share in the book.

Chapter 1

Great EHS Training Starts with Change

Imagine environmental, health, and safety (EHS) training that everyone wants to attend, your boss supports fully, and when you're finished leaves you feeling like you made a difference in the lives of the learners and to the business's bottom line. Doesn't that sound like a great way to keep people safe? Wouldn't it be awesome if that's what happened every time you trained? But that's not what EHS professionals generally feel when it comes to training. Instead, we are faced with a myriad of reasons for training either failing outright or not really helping to save lives or to make a better workplace. Challenges to great training include many of the following:

- not enough time to train and do everything else we need to do
- not enough time to deliver great materials
- a budget too small to use the best equipment, materials, or people
- EHS professionals who have never been taught what great training really looks like
- training topics driven exclusively by reaction to accidents, complaints, or legal trouble
- topics dictated by regulations
- lack of support by supervisors and managers to release learners from regular work to be trained
- supervisors who don't attend the training their staff attends
- poor training rooms or facilities that lead learners to think the training is unimportant
- no budget for food or treats during training

- trainers who lack the technical knowledge to feel comfortable training
- subject matter experts who just read from slides

EHS training has a bad reputation. So bad, in fact, that EHS training is sometimes seen as punishment. The entire industry needs to stand up and take ownership of this problem. We have great opportunities to improve safety training as the tool that helps save lives, engages learners, and increases company profitability. But we need a shift in how we design, develop, and deliver our training. With that shift comes a change in the behavior and knowledge of EHS professionals. Great training delivery needs to be prioritized to the same skill proficiency level as regulatory knowledge, leadership skills, and risk assessment. If you are reading this book, I believe you are ready for this challenge. Know that I do not blame EHS professionals for this problem. It's not our fault that no one ever showed us another way to train. Our profession has passed on poor training skills to each successive generation. But we can change and improve with the right tools.

With all these challenges in front of us, it almost seems impossible to envision that great, effective training even exists, but it does. This book won't fix budget challenges or make your training space bigger, but it can help you create valuable, effective training that you can demonstrate to learners, supervisors, and stakeholders. It can help you refine your design, development, and delivery skills, and that will convert nonsupporters into your best advocates. There are systems, tools, and education that can make you a great trainer, and, better yet, your learners will thank you for training. Yes, they will actually thank you for safety training!

The Road to Change

If you think back to the first training you received, perhaps even the first program you delivered, was it much different from today's training? Maybe you have made some significant improvements, which is great. But many of the learners I teach about teaching admit they can easily fall into the trap of death by PowerPoint. One of the greatest teaching tools we have has been weaponized against the learner. But there are solutions, and this book will cover many of them.

Face-to-face training (FTF) is the gold standard for training delivery. Despite advances in e-learning, most EHS professionals feel that FTF is the best delivery method for us. This may be true in many cases, but with corporate globalization and remote workers, we have to acknowledge other options as either the most cost-effective or the best for some learners. There is an entire chapter on e-learning later in the book that may help you think of ways to use technology to make your job easier—and maybe even more fun! Part of what you will learn in the book is how you need to consider the audience when designing training. What you will read may help you see that the old way may not always be the best way.

Compliance training is the backbone of most of the work we do. We cannot simply choose not to deliver what is legally required, but we can start to look at other ways of transferring knowledge to the learner than the typical FTF death-by-PowerPoint option. How we design our materials, what we include, and how we can make learning more interactive—even when it is compliance training—are the first bridges to cross on our road to change.

Boring EHS training has become its own meme. I introduce myself to learners, and their reactions are priceless. “Safety training? We have to do more safety training?” I have always chuckled at evaluations when the learner writes in the comments, “First safety training class I have been to where I wasn’t bored!” Boredom is one of our biggest roadblocks to success. The tools, systems, and ninja tips in this book can help you immediately transform EHS training from boring to engaging.

Lengthy training is another challenge we face. Most of the learners are used to moving all day, not sitting in a chair. Even the best training program is tough to love when it’s an all-day event. Creative use of activities, breaks, and interactions are some of the ways to make full-day training valuable. An even better solution is to evaluate what you are delivering to learners and determine if you are teaching unnecessary materials. If you have to cover everything, consider breaking the training up into two half-days. This may make the production and operations staff happy too.

The dreaded lecture also continues to give EHS training a bad reputation. All the limitations above often force us to use the same PowerPoint presentation we have used for years. Whatever your reason for recycling poor training materials, now is the time to commit to a new, improved program. Although some of the up-front work will take more time, the end result will be better for you and your learners.

Why Become a Safety Training Ninja?

Because you and your learners deserve it! You deserve to work smarter, not harder. You deserve to be recognized for the hard work you do. You deserve to focus your training time on ensuring you are really helping learners work more safely. And you deserve the credit when the training boosts production and quality, and, best of all, saves lives!

Your learners deserve to have their time and needs respected. Delivering the training they need—exactly and only what they need—is a great way to show you respect them as individuals. They will want to come to EHS training when it is interesting, valuable, applicable, and succinct. If your learners are in a high-production environment, every minute in training is a minute away from making a deadline or quota. They may really want to be safe, but it's impossible for them to forget what else they need to be doing.

And the industry deserves it. Sadly, EHS training is viewed so poorly that many professionals don't want to do it, so it's delegated to someone who may not know what they are doing. There are a few of us out there begging to fix broken EHS training, but many in the industry struggle with success. It is clearly needed, and the American Society of Safety Professionals (ASSP) and the Board of Certified Safety Professionals (BCSP) have even elevated the issue. ANSI/ASSP Z490.1, *Criteria for Accepted Practices in Safety, Health and Environmental Training*, was first published by ASSP in 2009 and revised in 2016. The BCSP began administering the Certified Environmental, Safety and Health Trainer certification in 2012. When you use the solutions provided in this book, you will be aligning your EHS training to both industry touchpoints. Part of the discussions you can have with stakeholders about budget and change can be justified with these alignments.

ANSI/ASSP Z490.1

To foster that alignment, I will be referencing and comparing ANSI/ASSP Z490.1, *Criteria for Accepted Practices in Safety, Health and Environmental Training*, throughout this book. Despite EHS training being a component of almost every EHS professional's scope of work responsibility, relatively few EHS pros own a copy of Z490.1. I strongly suggest you get a copy for yourself from ASSP.

Z490.1 specifically states the need for management of a comprehensive training program with elements that include:

- responsibility and accountability for the training program;
- resources available to the trainer and trainees;
- delivery strategy(ies) appropriate and effective for the learning objectives;
- appropriate evaluation strategy(ies) included in all training; and
- a system to evaluate the overall quality of the program managed to ensure consistency and continuous improvement. (ANSI/ASSP Z490.1, 11)

Certified Environmental, Safety and Health Trainer (CET)

This is the only certification for EHS trainers with experience and expertise in developing, designing, and delivering environmental, safety, and health training. The CET's existence is an indicator of the importance of great training. No other specific discipline within the EHS industry holds its own unique certification. Although there are others for industry expertise and even management, the CET is special in that it focuses only on the issues of training. To earn the CET, EHS professionals must demonstrate a knowledge of the theory and practice of basic adult education while confirming knowledge and experience in the EHS specialty area(s) in which they teach.

Specifically, professionals must:

- document 135 delivery hours of teaching or training in any EHS specialty
- hold one of the following BCSP-approved SH&E credentials (for a full list, see BCSP.org):
    - Associate Safety Professional (ASP)
    - Certified Dangerous Goods Professional (CDGP)
    - Certified Fire Protection Specialist (CFPS)
    - Certified Hazardous Materials Manager (CHMM)
    - Construction Health and Safety Technician (CHST)
    - Certified Industrial Hygienist (CIH)

- Chartered Member of Institute of Occupational Safety and Health (CMIOSH)
- Canadian Registered Safety Professional (CRSP)
- Certified Safety Professional (CSP)
- Occupational Hygiene and Safety Technician (OHST)
- Safety Management Specialist (SMS)
- Safety Trained Supervisor (STS)
- Safety Trained Supervisor Construction (STSC)
- NEBOSH—National or International Diploma in Occupational Health and Safety (BCSP n.d.)

I have long advocated the pursuit of the CET in our profession. When I teach Safety Training Ninja classes and seminars, I take the time to promote the importance of the CET. Expanding the number of certificants will further support the delivery of great training to our learners and promote the business objectives of our stakeholders. For more information or to start your application for the CET, go to BCSP.org.

45001

On the international level, in 2018 the International Standards Organization (ISO) completed ISO 45001, *Occupational Health and Safety Management Systems—Requirements with Guidance for Use*. Shortly after its introduction, the United States adopted it as a national standard (ANSI/ASSP/ISO 45001-2018). ISO 45001 mentions training more than twenty times as part of effective implementation of the guidance.

education theory

If you choose to pursue the CET or other educational certifications, you will need to know about many of the terms listed in table 1.1. As you continue through the book, try to remember that many ninja skills are grounded in strong adult education theory and using them will make you a better trainer and more likely to be remembered.

Safety Training Ninja Terminology

We'll be referencing specific terms throughout this book. They are listed in table 1.1. You may be familiar with some of these, while others will help you on the road to becoming a Safety Training Ninja.

TABLE 1.1 Glossary of Terms

Term	Ninja Definition
ADDIE	A cyclical instruction design model that has five phases: Analyze: Gathering information that determines how to bridge a learning, knowledge, or skill gap Design: Planning to achieve your learning objectives Develop: Creating the materials for your training program Implement: Delivering your training program Evaluate: Improving your training materials based on feedback and the experience during course delivery
adult learning principles (ALP)	A collection of theories and concepts documenting how adults learn and how it is different from how children learn.
andragogy	The term used to describe adult learning, based on the writings of Malcolm Knowles.
asynchronous training/learning	Used in the e-learning realm, the term applies to learning that the learner completes without the facilitator present. Examples include, but are not limited to, on-demand video, online interactive training course, or previously recorded webinar.
attendee	An individual participating in a safety conference, class, or convention; also known as a learner or delegate.
blended learning	The concept of combining different types of learning delivery systems to achieve your learning objectives.
Bloom's Taxonomy	Based on the work of Benjamin Bloom and later revised, the learning outcomes are classified in seven sections from simple remembrance to complex evaluation.
brainstorming	An all-ideas-welcome process used by a group to generate ideas to improve a process, skill, or knowledge.
case study	An incident or example used to illustrate the details of an occurrence. Often performed in groups during training with the goal of improved performance, knowledge, or skills. Great for hazard control solutions.
chunking	A process of grouping lists or concepts into seven, plus or minus two, sets to increase memory retention. It is your job as the designer and facilitator to determine when your materials should be chunked for better learning.
class	The group of learners you are working with. Could be five people who are watching some OJT on a new machine or an official roster of learners carefully curated by your stakeholder.
class culture	The collective attitude of a class or training group as reflected by their behaviors.
classroom	This is wherever you are presenting your materials. For the EHS trainers, this could be an auditorium, classroom, on the shop floor, or leaning against a piece of heavy equipment. Don't get hung up on the word *classroom*—it means wherever you will be with your learners.

TABLE 1.1 Glossary of Terms (cont.)

Term	Ninja Definition
compliance training	Mandatory training requirements as dictated by law, regulation, or consensus body.
computer-based training (CBT)	The use of computers to deliver either synchronous or asynchronous training.
delivery	A system of presenting information to learners; for example, instructor-led lectures, books, via a learning management system (LMS), etc. It is the implementation phase of the ADDIE cycle.
e-learning (electronic learning)	An all-inclusive term used to describe online training, web-based training, webinars, videos, online classrooms and working groups, or any other internet-based training.
evaluation	The method or system of analyzing the effectiveness and impact of training.
face-to-face (FTF) training	Where the learners can see the trainer. This historically occurs in the same location but can now take place in remote locations thanks to video conferencing and camera-sharing technology.
facilitator	The role a trainer plays when efforts are more focused on guiding and coaching, not lecturing.
feedback	The relaying of information on the success or improvement opportunities of a training program; this can occur formally during the evaluation stage or informally through conversation during and after a training event.
Gagne, Robert	A psychologist best known for his nine steps of instruction; his concepts are part of mastering adult learning principles.
goal	The desired state of knowledge, skill, or business impact after the training is complete; this can include both business and training goals.
icebreakers	Energizing activities conducted during training, usually at the beginning of the program or each day for multi-day programs. Design icebreakers to put learners at ease with the content, trainer, and each other.
informal learning	Learning that takes place outside of the formal training program; examples include coaching, content-based conversations, job aids, and nonstructured on-the-job training. Recent additions to the concept include self-directed readings, practice of skills, and peer-to-peer learning.
instructional designer	The person who uses ADDIE or another instruction design system to design and develop training materials and their supporting content.
instructional systems design (ISD)	The system used to create training. ADDIE is the original system. Others now adopted include SAM (successive approximation model) or AGILE (Align, Get set, Iterate and implement, Leverage, and Evaluate)
instructor-led training (ILT)	Traditional training, which we use a lot in EHS. This is often used interchangeably with FTF. Usually considered synchronous, onsite, or sometimes using web conferencing technology.
job aids	Tools provided during training or after to support the learning objectives. Valuable for learners if the materials are complex, checklist based, lengthy and hard to memorize quickly, or only performed periodically and easier to forget due to lack of repetition.
KISS	It means "Keep It Simple, Silly." You may have heard it phrased other ways, but whatever the phrasing, we need to remember to keep our training simple and to the point.

TABLE 1.1 Glossary of Terms (cont.)

Term	Ninja Definition
Kirkpatrick, Donald L.	The founding father of training evaluation. The Kirkpatrick Model of evaluation includes these levels: Reactions, Learning, Behaviors, and Results.
Knowles, Malcolm	The father of adult learning theory, who defined the six assumptions of adult learning. This theory relates more to how you need to communicate respectfully with people than to a prescription of delivery.
learning	Gaining knowledge, skills, or understanding through training, experience, or self-discovery.
learning management system (LMS)	Software that manages information about the training you provide. It can host e-learning as well as track FTF training. It can provide reports, evaluations, and test scores on learners and courses in more automated systems.
learning objective (LO)	Learning objectives define what your learners will be able to do at the end of the training or your stated timeframe. Your learning objectives will describe who will do what, how well, and by when.
learning style	How the learner accepts and retains information presented; the style is influenced by how your learner behaves, reacts, feels, and processes the learning.
Likert Scale	The response scale used in most reaction-type evaluations. The scale allows learners to rate the level of agreement or satisfaction to questions. The Likert Scale is usually based on a 1-to-5 or 1-to-10 rating.
microlearning	A concept usually associated with teaching and delivering learning materials in small and specific bursts of content. Learners usually control where and when they learn. The technology-savvy learner often easily adopts this useful option when provided electronically.
mnemonic device	A memory-triggering tool usually connecting material to be learned/remembered to something more meaningful to the learner.
observation	When a learner watches others to increase skills or knowledge; can also be used post training to confirm learning retention and application.
on-the-job training (OJT)	May be formally or informally delivered. OJT happens through observation, peer-to-peer training, and leader-to-learner training. It often occurs spontaneously and therefore may miss some critical elements for good training and may not be recorded as actual training in the LMS or learner performance records.
pedagogy	The opposite of andragogy; relating to the practice of teaching and how to influence students. It is the learning method typically used in children's education.
Phillips, Jack	Leading expert on return on investment (ROI). He added a fifth element to Kirkpatrick's four levels of evaluation to include ROI on training programs.
project management	The planning, implementation, and delivery of training programs; may include management of resources, time, budget, improvement strategies, and learner outcomes.
role play	An energizing and exciting activity often used during training to assist learners in practicing skills or developing communication capabilities.
smile sheet	The alternative name for a level-one evaluation of training.
social media learning	Leveraging social media as a tool for learning. Using Twitter, Facebook, Yammer, and others to advertise, promote, affirm, and follow up on learning. Can also be used independently for microlearning.

TABLE 1.1 Glossary of Terms (cont.)

Term	Ninja Definition
stickiness	How well your training message "sticks" with the learners.
storytelling	The weaving of relevant and impactful stories into the training to connect with learners and offer an alternative way to recall information delivered.
subject matter expert (SME)	A person with expert knowledge on a topic or process that can assist in the ADDIE cycle.
synchronous training	When learners and trainer participate at the same time; can be FTF or remote, using technology
teach back	When learners deliver materials learned back to the other learners, usually in FTF training.
Thorndike, Edward Lee	Adult education pioneer whose laws and theories date back to the early 1900s.
trainer	The person who delivers learning to improve performance skills or knowledge.
transfer of learning	The stickiness of learning, when materials are learned, retained, and then applied or implemented on the job.
video-based learning	Using high- and low-tech video to teach specific behaviors or skills that can best be demonstrated via video.
web-based training (WBT)	A synonym for e-learning or other training that is delivered via the internet.
WIIFM	This stands for What's In It For Me? WIIFM is a reminder that we need to focus on the audience's needs, not the trainer's needs.

References

ANSI/ASSP Z490.1-2016. 2016. *Criteria for Accepted Practices in Safety, Health and Environmental Training*. Park Ridge, IL: American Society of Safety Professionals.

Board of Certified Safety Professionals (BCSP). n.d. Certified Environmental, Safety and Health Trainer. Accessed August 7, 2018. https://www.bcsp.org/CET.

Chapter 2
ADDIE: Using a Systems Approach to Training

How to Deliver Better Training

In the previous chapter, we talked about the challenges we face. Safety is a full-time job for most of us. Add to that the responsibility to train effectively and efficiently, and that is probably more work than we have time for. So let's find a better and easier way to train.

We use a systems approach with several safety initiatives, so it makes sense that we can adopt a systems approach for training. If you are an inquisitive type who has started to research good training design and development, you may have seen the terms instructional systems design (ISD) or systems approach to training (SAT). No matter which acronym you choose, SAT or ISD, you are basically talking about the same concept. By the way, I will be using ISD since that is what I first learned.

Although it existed to some degree before the 1970s, the ISD concept was really born at Florida State University and grown by the US military shortly thereafter (Branson et al. 1975). Of the different ISD models, we will focus on the ADDIE model, which stands for Analyze, Design, Develop, Implement, and Evaluate (see figure 2.1). When I was a younger Safety Training Ninja, I learned the ADDIE model, and although there are other, newer rapid-development programs, I think we should start with the basics to grow our skills.

Many of you may already be using some of the elements of ADDIE but didn't know its official name or that it is an accepted ISD model. ADDIE

FIGURE 2.1
The ADDIE cycle

ANALYZE

The analysis phase is the foundation of a training program. This is where some pre-planning will serve you well. The basis for who must be trained, what must be trained, when training will occur, and where the training will take place is completed in this phase.

DESIGN

This is the road map for your training program. The outputs of the analysis phase drive the design and end in an outline of the training program for future development.

DEVELOP

This phase begins the creation of your program. We start to expand on the objectives and start filling in the outline. The "filling" is both the technical content and the methods of delivery.

IMPLEMENT

This is it, time to teach, train or facilitate. All your hard work and materials preparation are finally ready to deliver. If you mapped out the design based on your analysis, delivery will go a lot easier.

EVALUATE

The last phase is to evaluate if you met your learning objectives and the materials you delivered will fix that knowledge or performance gap you found in the analysis phase.

is orderly—think military—and that is a good thing. In the safety world, we need to think things through, and training needs the same thoughtful approach. The key to ADDIE is using it to your advantage, not thinking of it as a stiff compliance approach that can't bend to your needs. ADDIE allows for continuous improvement throughout the process. That is valuable since we use continuous improvement in many other approaches to EHS, such as:

- Plan, Do, Check, Act (PDCA) in ISO 45001 and quality processes
- OSHA's Process Safety Management regulation
- MIL-STD-882D, *Department of Defense Standard Practice for System Safety*

Support for a Systems Approach

Who else uses systems for EHS training development? Plenty of people already, and you can join this esteemed group.

- ANSI/ASSP Z490.1-2016, *Criteria for Accepted Practices in Safety, Health and Environmental Training.* This is one of the few consensus standards specific to the EHS training world. Using this standard to complement the ADDIE approach will help you create and deliver better training. Following the project management elements of this standard will help you defend your training program contents.
- "Best Practices for Development, Delivery, and Evaluation of Susan Harwood Training Grants," US Occupational Safety and Health Administration, (https://www.osha.gov/dte/sharwood/best-practices.html). The Harwood grant program has been around for more than twenty years and has long advocated many of the elements of a systems approach to EHS training.
- U.S. Army Training and Doctrine Command (TRADOC) pamphlet 350-70-7, *Army Educational Processes* (don't be fooled, this pamphlet is 107 pages long!). If you work with or for the Army, you are probably familiar with this document. It is comprehensive, to say the least, and not an easy read, but it shows again that EHS training programs are being developed using a system. There are many more to list, but this should give you a good start to justify following these new concepts.

Why ADDIE?

Maybe you are not convinced that this systems approach to EHS training is for you. Consider what ultimate goals your training programs need to achieve. Do you want great training that learners want to attend and that they actually use at work to be safe? Then this might be a good start at achieving those goals. Here are some other benefits of ADDIE.

1. Break out of the mold. Use the ADDIE model to break the old mold of developing EHS training. You know the one: open PowerPoint, cut and paste all the stuff you can find in your files or on the internet, and then supplement with lots of regulatory content that makes even the most dedicated learners' eyes glaze over. By developing materials systematically, you focus on only what they need to know, whether by law or job necessity, and this allows your learners to hear only what they need to know to fix any knowledge or performance gap.
2. Only teach what they need. Once you know what the learners need to know via your analysis, you can start removing all the stuff you thought you had to include in EHS training but don't need to meet your learners' or stakeholder's needs. If the material you are teaching does not actually say they must be able to recite the regulation reference (29 CFR 1910—you know where I am going with this), or cite the scope of the regulation that does not apply to them, or list other elements of a rule or governmental requirement, then why would we include it in the content we teach our learners? Is it because we have always done it that way? Is it because that is the way we were trained? Or is it because we don't know another way? Well, keep reading and you will soon find a better way!
3. Stick to critical information. If you are conducting training because you have identified a specific performance or knowledge failure, then limit your training to improving that specific element. For example, an analysis of recent accidents has found several ladder accidents. Your further analysis has found that it is just one ladder in one part of your facility. Now you know to train only the learners who use just that ladder, and you can focus on the specific behavior you need to change. No need to reteach the unneeded elements of any government regulation. Many regulations require a proficiency to work safely, not the recitation of elements that do not apply to

ditching the regs

Not sure if you are ready to leave behind all those references to any US OSHA regulations? Refer to OSHA document #2254, *Training Requirements in OSHA Standards*, to decide just how specific you need to be in your training program.

the work. And don't forget how much your learners will appreciate training that is specific to their needs and no more.

4. Protect your professionalism. If you know this ninja, you know that I don't like to use regulations or litigation as the primary reason to be safe or teach learners about safety. Imagine if the quality or even the existence of your training program were to be challenged. If you are following a systems approach, with the project management paperwork behind it, you will be able to defend your training programs and show the "how and why" of what you trained your learners.

So, let's break the ADDIE phases down, just enough to get you thinking.

Analyze

Sometimes this is the hardest part of the effort for trainers: getting started. If you are sure of your analysis, you are more likely to proceed smoothly through your training development. We will talk about analysis in greater detail in the next chapter.

The analysis phase is the foundation of a training program. This is where some preplanning will serve you well. The basis for who must be trained, what must be trained, when training will occur, and where the training will take place is completed in this phase. The product of this phase is the foundation for the remaining elements of the ADDIE model. This is a critical stage since, as an EHS professional, you are probably pushed to get the training developed quickly and cheaply. The problem is, if you skip this step, you may actually end up wasting time and money. It may take some convincing for both you and your stakeholders to invest in this phase, but I promise it is worth it!

Design

This is the roadmap for your training program. The outputs of the analysis phase drive the writing of learning objectives, which drives design, which all results in a road map of the training program for future development. This is my favorite phase because it allows the EHS professional to finally break free of lectures and less-than-stellar training. The other great thing is, if you did your analysis well, the design process starts to form organically in your mind.

Develop

This phase begins the creation of your program. We start to expand on the objectives and start filling in the road map. The "filling" is both the technical content and the methods of delivery. The result is the completed instructional materials. Don't get ahead of yourself, though; the planning and the road map must be developed first. Developing different forms of course materials requires a certain amount of skill and art. Be ready to acknowledge that you may not have those skill sets, but you can start learning now.

Implement

This is it, time to teach, train, or facilitate. All your hard work and preparation of materials are finally ready to deliver. The implementation has far more opportunities than many EHS professionals take advantage of, and if you map them out properly in the design and development stages, the delivery will go a lot smoother. The chapter on implementation will have both technical and artistic information to help you deliver and facilitate the materials you worked so hard to create.

Evaluate

The last phase is to evaluate whether you have met your learning objectives and if the materials you delivered will fix the knowledge or performance gap you found in the analysis phase. Testing, job evaluation, employee observation, and loss trend analysis (after time has lapsed) are all great tools, and we will expand on these concepts well beyond the pre- and posttests that are common in the EHS training world.

References

Branson, R. K., G. T. Rayner, J. L. Cox, J. P. Furman, F. J. King, and W. H. Hannum. 1975. *Interservice Procedures for Instructional Systems Development* (5 vols.) TRADOC 350-30. Ft. Monroe, VA: U.S. Army Training and Doctrine Command, August 1975.

Chapter 3

Analyze Your Audience and Your Business

The goal of a needs assessment, or the analysis phase of ADDIE, is to determine the current and the desired levels of knowledge or performance. The gap between the two is the learning that must occur and is the basis for a good training program. I wrote before about finding exactly what your learners need to know and creating a program that meets those and only those needs. This gap may be an actual problem that you discovered doing your daily EHS duties. Perhaps someone mentioned to you they have a safety concern and want to learn more, or you were conducting an investigation and recognized an accident trend that training can fix. Sometimes it may not be so easy to identify the problem, much less solve it easily. Has a stakeholder ever said that you need to conduct training because a seemingly unlinked set of accidents or near misses have been happening? How do you know what to train when you are told to "make everyone work safely"?

Here are some sources of information or behavior triggers to help find gaps in knowledge or performance:

- supervisor or management comments and feedback on what they see and experience in their work
- reports from safety audits, safety reviews, and quality and production reports as they may relate to EHS
- equipment work orders and maintenance requests
- worker complaints, comments, and suggestions
- changes in work, location, cell, process, or operations
- changes in production schedules, turnarounds, or shutdowns

- additions of new equipment or processes
- changes in job responsibilities
- policy or procedure changes
- accident and near miss investigations
- legal or regulatory change

What to Analyze and How to Make Sense of It

Once you have gathered information about the gap you are trying to bridge, you can start the analysis. This is where analysis gets valuable—it will help you solve the mystery of "how to fix that problem."

The analysis should help you answer the following questions:

- Will any training fix the problem?
- Will EHS training fix the problem, or is it another issue?
- Why wasn't it solved before if we have trained the materials before?
- Will the training solve our business needs? (More on this to follow.)
- Who really needs the training?
- What other performance issues will need to be addressed to ensure the training is effective?

Ways to Conduct Analysis

There are many ways to try to find out how to bridge the gap you have identified (or are still trying to identify), even more ways than we will talk about here. Initially, I will focus on some of the ways you can gather information that you will be able to use right away. Of course, time, money, location, and your safety and training culture will all impact how accurate and complete the analysis can be. But even with almost no budget and a remote workforce, you can find out more about the knowledge or performance gap of your learners than you may have known before.

Interview or Conversation

Have a conversation with the learners and their managers and anyone else who might know about the identified problem you think training will fix. If you are lucky and you can walk right up to the learners and ask them, you will likely get some really useful information. But what if you can't do that?

Try to find anyone who knows about the work you are analyzing. A phone call will work if you can't talk face to face. Missing this source of analysis will often result in the training missing the mark. Do your best to gather as much as you can, despite the challenges you may face.

Watch or Observe the Work

Be thoughtful with this process. Explain why you want to see their work: to help you help them and others work more safely. As EHS professionals, we need to take that extra time to let the learners know that this is to help us do our jobs better and not to catch them doing something wrong. I have found that an open and honest approach to learning more about how someone does his job is a great way to learn about how to do it more safely. They may even have suggestions on production and quality. This part of analysis is also important because it may help identify other constraints that could prevent effective training. As a bonus, the learners who help you in this process can often be supportive during the implementation of the training because they were part of the solution.

Review New Law or Regulation Thoroughly

If there is a new law or regulation, you need to understand it before determining what your learners need to know. A good ninja does not assume you need to train on the new rule, regardless of the requirement. You may already be doing everything you need to be doing to comply. You won't know until you conduct your analysis of the requirements and then compare that to the way your learners are working. You might only need to provide some new elements, or you may have a lot of work to do. Avoid the reflex to go online and download some version of the new rule's training program developed by gosh-knows-who. Instead, do a little up-front work to avoid bad training or wasting your and the learners' time.

Conduct Accident and Near Miss Reviews

Dig into the data you have to help you improve learner safety. If you work for a large organization, you may be able to spot trends in losses and near misses that can really make a difference, especially if no one else is looking. Analysis of losses for training can also help you pinpoint the behavior you wish to change or improve. Consider this example. Recently there have been three near misses involving forklifts and pedestrians at your plant. You conduct an evaluation of the near miss data and figure out that all the incidents occurred during third shift—different drivers, but only on one shift. You see forklift operations on all three shifts and find that third-shift operators are running the equipment differently than the other shifts. Now you can focus your training on just the operators who need to improve their safe operating skills instead of everyone at the plant. This kind of analysis may not be new to you, but you may not have realized it was part of a bigger EHS training system such as ADDIE. You may have been using ninja techniques already, but you just may not have known what they were called.

In ANSI/ASSP Z490.1-2016, section 4.2.2 reminds us not to forget to consider additional elements in our analysis, such as a review of job analyses; any relevant site-specific information; and whether the learners have ability, language, culture, or literacy issues that can impact the success of your training program. Annex B, Training Course Development Guidelines, supports the gathering of information we have talked about already, including:

- job-specific task analysis
- learner interviews and observations
- learner questionnaires
- supervisor questionnaires
- management interviews
- regulation analysis
- incident analysis
- applicable regulations (ANSI/ASSP Z490.1-2016)

analysis help

Conducting your first analysis can be overwhelming. Reach out to potential learners who have shown an interest in EHS training in the past or who know something about your training topic. Ask for their help determining the audience, the knowledge or skill gap, and potential ways to make the training you are considering better.

Identify Your Audience

During your analysis of what the gap is, you will be defining the audience for the training by default, ensuring you train everyone who needs the information and only them. In some cases, you may train next-level managers if they also need to know how to do particular tasks or activities or to be sure they manage the learners in the best way possible. Include next-level management in any discussions about details of the need for the training, and they can help develop a list of exactly who on their team may need to attend. We will discuss the learner audience in more detail in another chapter. For now, keep in mind the need to determine who should attend the training as part of your analysis.

In my first year of training, just out of university, I was conducting respiratory fit testing for my organization's employees. I was new and did not design the course nor determine who came to the class. The system already in place had leadership notify managers of training programs their teams would need to attend, and the managers made sure employees attended those classes. Well, the flaws in that system quickly emerged when I looked around the class and offered up a treat for whoever had gone the longest since they had last shaved, especially for the fit test. A pink-faced gentleman raised his hand and said, "Twenty-seven years!" He, of course, won the treat. I then launched into the training of the why and who of respirators. He was kindly paying attention and asked if truck operators doing his type of work needed to wear respirators. He had never been required to wear one, and he had never seen anyone wearing one. I asked him some questions, and we determined he was not doing any work that would put him in the category of needing to wear a respirator . . . ever. His manager did not bother to determine if he needed to attend the training or if his job required the use of respiratory protection. It was a tough moment to recover from, but I did so by asking to talk to him during the break and then getting the class back on track. I never forgot the importance of knowing who should be trained and never letting anyone throw people into safety training to meet headcount requirements.

The organization had developed apathy for safety training, and I watched thirty workers scratch their heads about management's commitment to valuable training. The reputational damage to the EHS team was huge, the relationship with the supervisor was damaged, and I looked like a fool. Maybe we cannot always prevent errors such as this when we are told to show up and teach. But as the person responsible for the analysis and design of a program, you can take the time to be specific about exactly who should attend which training course.

Determine If Training Is the Solution to Your Problem

Since this is a book about how to be the best trainer you can be, we assume that training is the solution to your performance or knowledge gap. But what if it's not? Before you get into ninja training mode, consider that the problem could be due to dozens of other things: inadequate materials, tools, or workspace; unreliable equipment; unclear expectations; inappropriate consequences or incentives; lack of feedback or coaching; inappropriate job assignment; or others.

Consider these questions before you move forward on your training program.

Motivational issues to safe work

- Are the learners doing the job they were trained to do?
- Do your learners get clear/prompt feedback on safety performance errors?
- Are they punished for errors?
- Are your learners clear on the role they play as part of a team?
- Are they rewarded for known or unknown unsafe behavior?
- Do learners receive mixed messages about the value of safety? Is it top on the list on paper—but only if production goals don't get in the way?
- Is unsafe behavior easier or more desirable to the learners?

Environmental barriers to safe work

- Is there a new process, equipment, or system in place?
- Are parts, tools, supplies, and equipment readily available to do the job safely?
- Is the work site/facility adequate to do the job safely?
- Is there enough staff to do the job safely?
- Are the workflow and procedures clear?
- Does management or supervisory staff change often?
- Is there high turnover in other staff?

Skill and knowledge barriers to safe work

- Is the work too complex to be memorized by the learners?
- Are the steps to working safely poorly communicated?
- Are the job aids telling the learners the wrong way to do work, even to do it unsafely?
- Is the OJT given incorrectly, inconsistently, or unclearly?
- Is the task new, and have the learners ever been instructed on how to do their job correctly?

If you have found motivational or environmental barriers to working safely, then training may not fix your performance or skill gaps. You will need to solve those problems before you train learners on new skills or behaviors. It would be great if everything gets fixed at one time, but it's not likely. A ninja will be challenged at this point because your stakeholder may not want to hear about the other issues. You will need to enlighten them; otherwise, your training program will fail because it won't solve the root problems.

Linking Your Training to Your Business

How do you get your stakeholders to allow the time needed to conduct a proper needs analysis and all the remaining steps? You have to connect the value of the training to the business objectives of the company. In the simplest of terms, can you connect the problem you are trying to fix with a value to the company? Unfortunately, saving lives and the moral need to protect the safety of our learners is not enough in the business world to justify training. Heck, many companies wonder if they even must train to regulatory compliance since they may not be convinced that they will get caught or that someone may really get hurt. The challenge we face as EHS professionals is the same when we try to get time or a budget for training. Saul Carliner advises in his book *Training Design Basics* to pick one of these three business goals:

1. Generate revenue
2. Control expenses
3. Comply with regulations (Carliner 2015, 47)

Sounds easy, right? Except it's not. EHS professionals have been trying to use compliance as a reason to train for years. While it may get us some bodies in classrooms, it doesn't get us the actual time and money we need to do it right the first time. So, you will need to dig deeper. The first try will be hard, but if it works, you now have a successful model for future projects. First, you need to consult with a mentor or someone in finance, or use your past knowledge of working with the C-suite to determine what the top priority of the business is. Is the company in revenue mode or trying to control costs? If you go the route of compliance, you will still need to tie the completed training to the other goals when compliance alone has not been effective in the past.

After you have completed the analysis and are starting to get a sense of what that gap is, try to decide if the completed training will help learners work better, safer, and/or with higher quality or production. Alternatively, can you show that the problem you are trying to fix has cost the company money in workers' comp losses, production shutdowns, or time and money in product reworks? If the answer is yes to any of these, then you can try to extrapolate some money made on faster production or money you can save on loss prevention. This is not a book on how to run those numbers, so again I advise finding a mentor or finance person to help if you can. In today's EHS profession, many of us have to figure out how to tie safety to profit or loss control for almost all we need to do—this is just an extension of that process.

Document What You Are Doing

You have been looking at data, asking questions, making observations, and conducting an analysis of learning needs. You have been doing good work, and now it's time to remember to document what you are doing. The *why* of your training project will help you build that business case and will help justify the work you will be doing. If this is your first training project using all the ADDIE elements, your notes and data will help justify future efforts. For the first ADDIE phase, ANSI/ASSP Z490.1, section 7.2, recommends a written project management framework that includes the following elements:

- target audience
- learning objectives
- training material content sources
- the actual training materials
- continuous improvement and evaluation components

Add your analysis of the gap to that, plus any financials or regulatory evaluations you can use or may want to go back to later.

Ninja Q&A

What if I don't have the time to do an analysis?

I know that conducting a good analysis takes time—a lot more time than I have described here. We didn't even go into the other ways to conduct a good analysis. But doing something, anything, to try to determine why you need to deliver the training, why it will fix a problem you have, and who should attend is the least we can do. Even if you don't call it an analysis to your stakeholder, you can have an informal conversation, send a simple email request to learn more about the problem, or utilize someone who knows about the work. Despite all the information on conducting an analysis, it is likely one of the first components to get chopped from the busy schedule of the EHS professional. But take the time to be sure you have done something to make sure your training is the right training.

What if I won't know my audience or their gaps before I arrive to deliver the training?

Hopefully someone did their job and designed the program after analyzing the gap to fix and the learners to attend. But that may not have happened, and if it didn't, a good ninja tries a few things to deliver the best implementation in a tough situation.

don't die

The ADDIE process will take some time, and often the EHS professional will have to shorten the first two steps. When I say shorten, I don't mean skip. I often remind learners that if you skip the A and the first D in ADDIE, it doesn't spell ADDIE anymore . . . it spells DIE. The whole point of being a training ninja is so your training program doesn't die in front of the learners!

- Ask the stakeholder for a list of learners and their job titles in advance—this is not perfect, but it's better than nothing.
- Make time during the opening to conduct an icebreaker where you can find out more about your learners. I demand my stakeholders allow me this time. I can't determine who they are and why they are there without it. It also serves to manage learner expectations if they are wondering why they are even in the class.
- Use the time before class to get to know anyone who arrives early. This serves many purposes and will help you know your class better.
- Use the learning objectives to really drive home the reason for the class. If someone is there by mistake, they might let you know during the break, and you can work on a solution at that point.

How long does great training take to develop?

Now I thought long and hard about whether I should include this, and you need to decide if you are going to share this with your stakeholders. Great training is a lot of work, and I am sure you know that. Below is the result of a survey conducted to determine how much work it really is. In 2010, The Chapman Alliance conducted a survey of 249 companies/organizations, representing 3,947 learning development professionals who have created content consumed by 19,875,946 learners. Those are some serious numbers, but it gets better. For instructor-led-only training (e-learning is another set of data), including front-end analysis, design, lesson plans, handouts, workbooks, PowerPoint, and SME reviews of content to be used during live, face-to-face learning events, the estimates in table 3.1 were generated.

It's likely you will get faster at your training program creation as you develop your ADDIE skills; conversely, the more you learn about what you can do to make your program better, you may take the technical work of program development to the next level: adding time. Will your work follow the above estimates or is yours different? Have you tracked how long it takes? The administration of training alone is a lot more time-consuming than many nontrainers realize, so could you be underestimating your efforts? Using table 3.1, when you can find the time, estimate what program development really takes and be honest and include *all* the work you do, not just PowerPoint and delivery time.

TABLE 3.1 Creating Learning

Number of Hours for Every 60 Minutes of Instruction	Complexity Level of Training Program Materials
22	ILT training, simple learning content, possible repurposing of learning source material with minimal learning support materials
43	ILT training, average project for creating corporate ILT class with well-documented deliverables (lesson plan, handouts, workbooks, PowerPoint visuals)
82	ILT training, complex subject matter, very customized, extended time spent on formatting classroom deliverables

Source: Chapman 2010.

Ninja Style: Lead Your New Training Culture

A good ninja knows when you are faced with a challenge, and changing the company culture about EHS training isn't going to be any easier than changing the culture in other parts of a company. But that doesn't mean you cannot be the leader of the new culture. It will be easier to start changing the culture when you can demonstrate the cost savings that come with the new culture. Your desire to deliver great, inspired, passionate training that saves lives might be hard to measure with dollars, but it's not hard to see. If you embrace what you are learning and add it to the passion I know you have, then you are closer to getting stakeholder buy-in.

For example, when you go to your executive team and explain that you need time in your schedule to streamline the annual hazardous chemical training to cut the instruction time and still achieve all the learning objectives and governmental requirements, I bet you'll get their support. To determine savings for the company, calculate how many fewer hours of training will be required for the learners, then multiply that by their hourly wage and the number of learners. Now maybe bump up the savings amount by a percentage point because shortened training time will mean more production time for employees. Throw in the bonus that the training will be better because you are adopting a training system used globally by the biggest and brightest companies and the US military, and you are on your way to generating some stakeholder support.

References

ANSI/ASSP Z490.1-2016. 2016. *Criteria for Accepted Practices in Safety, Health and Environmental Training*. Park Ridge, IL: American Society of Safety Professionals.

Bloom, Benjamin S. 1956. *Taxonomy of Educational Objectives, Book 1: Cognitive Domain*. New York: David McKay.

Carliner, Saul. 2015. *Training Design Basics*, Second edition. Alexandria, VA: Association for Talent Development.

Chapman, Bryan. 2010. "How Long Does It Take to Create Learning?" Chapman Alliance, September 2010. www.chapmanalliance.com/howlong/.

Chapter 4
Design Your Program with a System in Mind

We can't design without a road map, and it's necessary to have an outline before filling in the blanks. Usually either the lack of time or the inability to map out a training program is the reason our training misses the mark. It's not right, boring, too long, too short, or perhaps something else that leads to a less-than-amazing training program. But if you are willing to learn a new way to design and develop your program, you may find a better and easier way to do your job. So, let's get organized.

It's Not about You—It's about Them!

Whenever I teach the course How to Become a Safety Training Ninja, one of the most critical learning points I work to ensure everyone learns is this: When it comes to developing and delivering training, *it's about the learners*. Always. Every time. No matter what. It's not about you, the trainer; it's about them. Seems simple, right? Everyone knows that. Unfortunately, that is not what I see when coaching others to become better trainers, and it's not what I hear when I ask you or your learners about the training. The truth is, it's not that simple. But the good news is it's not completely your fault. We all have baggage, and sometimes that baggage carries over into our training program.

Here are some examples of how you might be putting yourself first, even if you don't mean to.

- You fight to keep things the way they have always been. As EHS professionals, we cannot accept that as an excuse for not being safe.

- You have large amounts of information on the slide and read from it. Yes, you know that's the wrong way to teach, but it's a governmental regulation. There's no other way to do it, right?
- You have more slides than you have time to cover, so you just skip over them and say they are not important or that you don't have time.
- You spend time teaching materials they don't need to know, but that makes you look really smart.
- You spend time teaching materials they don't need to know but that you want to teach because you think they're interesting.
- You have charts and data that no one can read—and you say exactly that when you teach—but you can't be bothered to change the slides so they are actually useful to the learner.
- You use cute pictures because without them the presentation would be boring. If the value of your program rests on whether you use clip art, you are reading the right book.

After just a 90-minute ninja training seminar, a learner sent me an email several days after the program and told me he had removed 40 percent of the content in the training program he was writing. He realized the content was not necessary to meet the business or learning objectives. All the extra materials were in there because of poor design habits, not learner needs. What was most rewarding to hear was how relieved the trainer felt. His training load got lighter, and for all the right reasons!

So, if you have been holding on to the old ways of training or you have a training program you have been using for years, step back from it and see if it's really the best choice for your audience. Do you have learning objectives? Do you know why you teach this course? Why must your audience attend it? Most important, does it fit with the business objectives of the company?

Creating Your Training Design Roadmap

There are many ways to deliver great training, and the safety world is starting to adopt more of those changes every day. For most of this book, I will be focusing on face-to-face (FTF) training with learners such as employees, clients, and workers. FTF has been the gold standard for training since the beginning of time, and it will always be my favorite to develop and deliver. Many companies and their EHS professionals are forced to train more creatively due to issues such as budget, travel, computer-based training, generational learning differences, and the international marketplace. Chapter 9 talks about e-learning in more detail and is a great resource for alternative means of training delivery.

There are many things to consider in the design stage. Your checklist may include: the materials, order, time available, ability to measure successfully absorbed material, the best way to approach different topics, and more. This section explains how to begin completing an outline using adult learning principles and different delivery methods, but first we must master writing learning objectives.

It's All about the Learning Objectives

Don't even think of opening PowerPoint yet! We have not talked about this before, but PowerPoint is closed. It is not yet time to open whatever software you are planning to use. We are close, though.

Before we start the brain dumping of all the information you know on the topic, I want to rein in that fast-moving safety mind of yours. The brain dump is not good for you or your learners. It makes a mess of your work and may trap you inside that mess instead of focusing on the problem the training is trying to solve and finding the best solution.

The key to your new, great design is to use what you learned in your analysis stage to create your outline. You create your outline based on well-conceived learning objectives. These are different than the business objectives we discussed before. Learning objectives (LOs) are what your learners will be able to do at the end of the training. For the EHS professional, this is usually a change in behavior or a change in knowledge of the learners. ANSI Z490.1, section 4.3.2, reminds us that proper learning objectives contain the following elements:

- the target audience
- the knowledge, skill, or attitude change we want to see
- any conditions or criteria relevant to measuring the success of the change (ANSI/ASSP Z490.1-2016, 16)

Although Z490.1 is right on the mark, let's take the ninja approach and simplify what is needed to write a good learning objective. Learning objectives must explain: *Who will do what*, *how well*, and *by when*.

learning objectives

The first attempt writing learning objectives is difficult for some trainers. Even if I am only teaching a one-hour ninja class, I have learners attempt to write at least one valuable learning objective for an upcoming program they are developing. That first try is critical, even if it needs some work. If you don't start trying to write valuable learning objectives now, it will only get harder later as you grow as a training professional.

For example, if you were delivering training on chemical hazards in a machining facility, you might have a learning objective that states: "By the end of this training program, machine operators will explain the hazards of the four coolant products used in this facility." *Who* is "machine operators," *will do what* is "explain the hazards of the four coolant products used in this facility," and *by when* is "by the end of this training program." The *how well* is wrapped up in the explain-the-hazards part of this example because it is something that can be tested, depending on what your program requires. This is a simple learning objective, but it's a good idea to start simple until you get good at it.

Getting SMARTer

You have probably heard of the learning mnemonic SMART. It stands for:

- Specific
- Measurable
- Attainable
- Relevant
- Time-bound

I like to use SMART to double check whether my objective is written properly. SMART is learning gold. If you don't get your learning objectives correct, you will likely miss the business goal of great training that provides real value (and that everyone wants to attend). If you want to develop and deliver great safety training that saves lives, saves money, and is actually good, you have to get the learning objectives right.

Bloom's Taxonomy

The first step to getting the objectives right is understanding Bloom's Taxonomy of Educational Objectives. Yep, you read that correctly. See table 4.1 for what is recognized as the comprehensive verb list of the cognitive learning process. Basically, Bloom categorized six levels of learning, and over the years, scholars and experts have expanded the concepts into action verbs and beyond. These

TABLE 4.1 Bloom's Taxonomy

Learning Category	Knowledge	Comprehension	Application	Analysis	Synthesis	Evaluation
Skill	Remember previously learned information.	Demonstrate an understanding of the facts.	Apply knowledge to actual situations.	Break down objects or ideas into simpler parts and find evidence to support generalizations.	Compile component ideas into a new whole or propose alternative solutions.	Make and defend judgments based on internal evidence or external criteria.
Verbs	Arrange Define Describe Duplicate Identify Label List Match Memorize Name Order Outline Recognize Recall Relate Repeat Reproduce Select State	Classify Convert Defend Describe Discuss Distinguish Estimate Explain Express Extend Generalize Give example(s) Identify Indicate Infer Locate Paraphrase Predict Recognize Review Rewrite Select Summarize Translate	Apply Change Choose Compute Demonstrate Discover Dramatize Employ Illustrate Interpret Manipulate Modify Operate Practice Predict Prepare Produce Relate Schedule Show Sketch Solve Use Write	Analyze Appraise Break down Calculate Categorize Compare Contrast Criticize Diagram Differentiate Discriminate Distinguish Examine Experiment Identify Illustrate Infer Model Outline Point out Question Relate Select Separate Subdivide Test	Arrange Asssemble Categorize Collect Combine Comply Compose Construct Create Design Develop Devise Explain Formulate Generate Plan Prepare Rearrange Reconstruct Relate Reorganize Revise Rewrite Set up Summarize Synthesize Tell Write	Appraise Argue Assess Attach Choose Compare Conclude Contrast Defend Describe Discriminate Estimate Evaluate Explain Judge Justify Interpret Predict Rate Relate Select Summarize Support Value

Source: Bloom 1956.

TABLE 4.2 Using Bloom in EHS Examples

Learning Outcome	Skill	EHS Example of the Skill Component of a Learning Objective
Knowledge	Define, describe, name, recall, identify, label, repeat	Name four chemicals used in your job.
Comprehension	Discuss, explain in your own words	Explain the hazards of working with acetone.
Application	Demonstrate or use knowledge in the appropriate situation	Locate any reproduction hazards listed in SDS.
Analysis	Analyze, break into steps, compare	Compare the PPE requirements between wood dust and pressure-treated wood dust.
Synthesis	Arrange, create, develop pieces into a whole	Develop the emergency evacuation plan for your work cell.
Evaluation	Appraise or evaluate based on knowledge	Rate the chemical inventory by most to least hazardous.

verbs are the key to writing your learning objectives (see table 4.2 for some examples). After almost thirty years of developing training, this ninja still goes back to Bloom when trying to ensure I write the best learning objectives. The material in table 4.1 is from Bloom's original work, which I prefer over the modifications in structure and concepts that other learning professionals have imposed over the years.

Let's practice writing a learning objective. Choose a behavior or knowledge change you want to achieve in a training program. Fill in the blanks:

_______ will ________ by ___________.

That second blank is where a good ninja leverages Bloom's Taxonomy, and it may be where you get a little tripped up on which words to use. Using your analysis of what the learners need to be able to do, find the action verb that best describes the new behavior or skill. Take your time to really look at the verb list, noticing how the concept of understanding is a category, *not* an action verb. This is a big deal, and you need to commit this to memory. A proper learning objective does not contain the word *understand*. Many EHS professionals have tried to argue this point, but trust this old ninja. Accept the fact that education professionals do not want you to use the word *understand* in your LOs and move on.

There are plenty of other great words to use. Think of the LO we used earlier: "By the end of this training program, machine operators will explain the hazards of the four coolant products used in this facility." *Explain* is the action verb. Refer to ANSI Z490.1, Annex B.8, for some additional details on what to ensure is in your learning objectives.

Why does all this matter? Because when it comes time to evaluate the training program and prove to your stakeholders that the training worked, you must be able to measure whether you achieved the learning objectives. I can measure *explain*. I can ask the operators to tell me about the hazards in a verbal evaluation, thus *explaining* the hazards. I have the option to write a test that requires them to explain on paper what the hazards are. I could even create a job aid for their managers to use to evaluate the learning by asking learners to explain the hazards to the managers. Ultimately, the changed behavior or skill of the learner is our desired outcome, and if we are successful, that gives us job security.

Your Learning Objectives Are Your Outline

If you can't write a proper learning objective, how will you teach one? Let's get back to ADDIE and how it can help us make your safety training great. If you have been able to write three or four great learning objectives that will lead to the achievement of the desired behavior or knowledge, then you have the beginnings of an outline. This is a budding ninja moment. Take note: Your learning objectives are your outline. Not that brain-dump mess from all the PowerPoints you can steal from the internet, not all the loose regulatory materials from any previous employees or friends in the industry, and not the words copied directly from any governmental or other mandatory rule. Don't worry; those things have a place—but not yet.

Embracing the outline based on your learning objectives is a big jump for most EHS professionals. This will be especially hard if you have already developed a program you think is pretty good, and now you are reading this book. It is also a challenge if you are using a program you did not develop but inherited. Many novice ninjas choose to apply the principles they have learned in my programs only to new training courses while still using old materials and methods for existing programs. Believe it or not, starting a new program is actually easier than revising an existing program. This may work best for you as well. In my experience, it is very difficult to turn your back on something

wait on creating the visuals

This is the time when many trainers will want to open PowerPoint or some other software and start cutting and pasting. Hold back on that action, place your learning objectives aside, and think. How can I teach the learners what they need? What concepts, examples, or activities would help me teach and make a great program. Jot those notes down. Often when I am teaching Safety Training Ninja classes, I find my learners start coming up with ideas during the LO development stage. This is good! Just don't get ahead of the process.

you worked hard to develop in the past or, worse yet, that your boss developed and thinks is the best training ever. Choose your battles carefully. I often advise other EHS professionals to start small and find a training program or topic that no one has an emotional attachment to. You know the ones.

Z490.1, section 4.3.3, states that the learning objectives should follow the SMART concept and should consider any required background and experience of the learners and knowledge or skill prerequisites (2016, 16). Make sure you've taken these into account in your outline and program.

Now you go back to your stakeholders with a plan to train on what is needed and only that. Take all the information you gathered using the techniques listed above and any others you may have and formulate a plan. Keeping the training to exactly what learners need to learn will keep your program on track and keep you from doing extra work.

Considering the Learners and Trainer During Design

Some of you will likely be training the same people many times in a year, while some of you may know nothing about the learners prior to arrival. When designing your course, you must consider the audience, either through analysis or assumption (based on the best analysis you can do). You need to find out as much as you can about the audience so that the course can match up to their needs as much as possible. Use this list to start thinking about them first and you second.

For learners:

- Are they interested in the subject? Should they be interested? How will you make them interested?
- What successes and issues have they experienced with safety training in general and with you as the trainer? How can this impact what you're trying to accomplish?
- What are their attitudes and beliefs relevant to the topic? If negative, how will you change that to positive the moment the class begins?
- Do they know why they are coming to training? Will you generate interest in the program before they arrive so you can be successful?
- Are their minds already made up? If yes, what will you do to change that before the class begins and then throughout the class?
- What are their opinions about you? If the way you deliver your training will change after you read this book, how will you prepare them for the change?
- How capable are learners regarding the training materials' content? Do they have the right knowledge and experience to learn what they need to learn?
- What are the learners' learning preferences and styles? There are many different learning styles and preferences; can you be flexible and build a program that allows for all or most of these styles? Can you facilitate a program that does all those things at one time?
- What are their attitudes toward training? What is the overall training culture? Is training seen as a perk or a punishment? Does it only happen when things go wrong, or is learner development part of a bigger organizational culture?
- How complex is the training? The more complex it is, the more likely it is that you will need to break it into chunks of data that are easy to learn. Are you implementing a complete change to a component of your safety management system? What barriers will be in your way that prevent the learners from wanting to learn?
- How much time do learners have to learn the new knowledge and skills? If you cannot train the learners in the time available, how will you change your course content to match the time? If you cannot shorten the content, how will you influence management to give you the time you need?

- How can you meet their needs? What are you going to do that is new or better to ensure that all future safety training will be best for the learners' needs?
- How will this training benefit the learners? Can you talk about the benefits of the training fluently? Can you discuss what they will learn and how it is not about "understanding" some safety- or compliance-related topic?
- Could there be disadvantages to the learners, such as production slowing down during learning? If yes, are you ready to acknowledge these and openly discuss the issues so they do not interfere with the learning opportunity?
- What changes do supervisors expect because of the training? Will you be able to deliver the change in behavior that you, the learners' supervisor, and the business expect? Can you prove it?
- Will the organization's culture encourage learners to use what they learn? How will you build the follow-up to the training? How will you make sure they are using the new knowledge or behaviors to improve the business? How can you avoid using formal or punitive audits to evaluate learning?

For the trainer:

- What is your style and technique as a trainer? Do you need to do your own professional development beyond this book to help you implement effective training? Are you open to delivering materials in a format other than lecture? Are you ready to consider that you may be part of your training problem and are not yet the solution?
- How complex is the training? Can you handle the content? Would it be wise to be more of a facilitator and bring in experts (even if they need delivery help too)?
- Do you have the technical experience needed to train effectively? Sometimes this works to your advantage if you can confidently facilitate a program when the learners know more than you do. When in that situation, I have explained to the learners that there is no way I can know more than they do about their jobs, but what I did know more about was helping them discover the safest way to do their work.
- Do you have the right facilitation experience? Working-group activities are hard to do! They're much more complex than we think they

are. When I coach budding ninjas, I remind them to set aside double the amount of time they think a facilitated activity will take. It is always better to finish a safety training course early rather than late.

- Are you ready for hostile learners? I was lucky—almost all of my first five years of training were in a hostile environment. You have to learn fast, be quick on your feet, and toughen your skin early in your career. I was always teaching new learners, and I knew their hostility wasn't personal, and that helped. But I still had to work hard to change their minds about compliance-based safety training.
- Do you bring your own baggage into the class? I was a woman in a male-dominated industry. Any baggage I brought because of that was my problem, and I needed to get over it quickly. If I was frazzled from the travel to the class, I needed to get over it. If the class the day before did not go as well as I wanted, I needed to get over it. If the learners learn, have a great experience, and see safety training in a new way, I get the credit for that. If they don't, well, I own that as well.

More questions to think about in the design stage when considering your learners:

- Will they ask many questions? This matters because it means you have built an environment open to the exchange of ideas. Congratulations! But it also means you have to build time in the instruction outline for the interaction.
- Do you expect them to raise objections? I hope so! It means they are engaged and interested. The best way to plan for handling objections is to anticipate what the learners may object to and practice your answers. If I know there will be major objections, I build an icebreaker around the objections. I usually tell them that this is their one chance to object or complain, and once they have said their peace, no more complaining is allowed. This gives everyone time to speak and then allows them to stop complaining. If needed, I will remind them of our agreement if they go off track later.
- Did the learners and their managers help identify the training needs? This is the key to your planning stage. Have you done an adequate analysis of the work, the training needs, and how you will deliver the message?

- What is unique about these learners? Upper-level managers may need a different delivery style than the laborers or first-level supervisors.
- Is there anything special about the location where you will be presenting? Will the location be its own distraction? If lunch is not catered, are there places to eat nearby that will not ruin your schedule? Is their adequate parking for the class? For you?
- Is there anything unusual about the date or timing of your training? Are you fighting against vacations, holidays, Mondays? Are you teaching during day shift and expect the night shift to stay awake? Is your training conflicting with anything else? Talk to Human Resources (HR) and the operations director about your timing and have them look at the calendar with you.

The above list is not comprehensive, but it takes all the complex, formal instructional design concepts and theories and simplifies them into questions that you can then use during your design stage. This self-reflection early in the process could help you discover opportunities for self-improvement and should make your job easier. I could tell you a story about each above bullet that I had to learn the hard way because no one told me.

Good ninjas do not think about themselves when they design, but about the learners. It is a challenge not everyone can meet. Ensuring the materials are accurate and helping learners achieve the learning objectives must be your main goal.

Icebreakers

Although some people may think that the purpose of icebreakers is merely fun (and many are), trainers use them for many reasons. Icebreakers may be used to:

- Generate interest in the class right from the beginning. We need that advantage in EHS training because of previous experiences of both the trainer and learners.
- Get learners involved in the class, which may be different from their past experiences.
- Provide nonthreatening introductions, which are valuable in EHS training, especially if the icebreaker concept is new.
- Set the tempo and tone for the learners. This lets them know that this class may not be like previous safety trainings.

- Make you a facilitator, not a lecturer, which is key to sharing the new ideas you learn in this book.
- Help you identify personality types; this can be especially helpful if you don't know your learners.
- Reduce your anxiety by creating a facilitated, relaxed environment that learners prefer.

Sequencing Materials

This often-overlooked gem will help you finalize your outline rather easily. The concept of material delivery sequence is very simple: it ensures your content is in the right delivery order (see table 4.3) and needs to be part of the final product. Go back to your learning objectives and make sure they are in order. Keep in mind, the sequencing can change within the program depending

TABLE 4.3 Sequencing Your Materials: Getting Your Order Right

Sequence Type	Use This When	Example
Chronological	Explaining the history or evolution of a topic.	The advancement of fall protection equipment and why we choose a particular piece of PPE.
Procedural	The order of the materials is dictated by something else, such as a law or policy.	Confined space entry training is a systematic approach to a highly procedural controls program.
Problem/solution	Your group must solve an EHS problem during the training course.	After a hazard hunt onsite, learners need to apply solutions based on hazard controls identified in the activity.
Categories	The topics are broken into different categories, and the course content does not depend on all of them.	Chemical hazard training—you don't have to teach the categories that your team does not need to know about.
General to specific	You must instruct on a broad topic of information initially, followed by specifics within that category.	Ergonomics is a broad field of study that may be taught generally and then specifically within subcategories that could include repetitive motion risks, soft tissue hazards, and controls.
Simple to complex	You have an increasingly complex topic. Start the instruction with simple concepts that build to more complex ones. (This reminds me of learning algebra in high school, each concept was the building block to the next, more complex application.)	Teaching the concepts of governmental compliance is more complicated than teaching the implementation of a safety management system.
Known to unknown	Start with a topic everyone knows to introduce comparison to an unknown topic.	This is particularly useful when changing a safety system or control. Explain the current system and how the new one differs.

on what you are teaching and the activities you implement. Check out the examples in table 4.3 that illustrate how to sequence in different ways.

Training Delivery Methods

Although we will discuss training delivery methods in greater detail in the next chapter, you cannot complete your training program outline without some thought about the delivery method you plan to use. Table 4.4 lists the pros and cons of six different delivery methods.

Training with Adults in Mind Using Adult Learning Principles

I know I promised to focus on those things you really needed to know to be a Safety Training Ninja, but indulge me in this section because it's important. Designing safety training with adult learning principles (ALP) is critical to success. As a matter of fact, a lot of the failing safety training I see doesn't consider it at all. A coincidence?

First, I want to mention what designing with adults in mind is *not*. It is not:

- the all-lecture format
- the cutting and pasting of safety regulations and rules on the screen
- gratuitous art that has no purpose
- flashy colors that only serve to make learners nauseous
- the "sit down and shut up" training course
- safety training as punishment for an accident or poor audit score
- the "I don't want to be here anymore than you do" training

There are many great masters of adult learning and instructional design out there. I want to focus on the three that get the most attention and make the most sense to this ninja.

Robert M. Gagne

Gagne was a psychologist and is best known for his nine events of instruction, listed below but tweaked a bit for simple explanation (Gagne and Medsker 1995). Consider these steps during the design, development, and implementation of your EHS training.

TABLE 4.4 Training Delivery Method Roundup

Delivery Method	Why	Why Not
Trainer-led presentation or lecture	• Keeps group together and on the same point • Allows good time control • Good for larger groups • Good for introducing new technical learning	• Learners find it painful and boring • Lends itself to being about the trainer not the learner • Is used too often in EHS training programs • Makes it hard to tell if people are learning • Provides limited retention
Working-group activities	• Provides better retention from thinking and doing • Allows practice of new skills in a controlled environment • Learners are actively involved • Allows for small group discussion	• Requires preparation time • May be difficult to tailor to all learners' situations • May be affected by different levels of management in the room • Needs sufficient class time for exercise completion and feedback
Reading assignments and individual exercises	• Saves time (learners can read faster than trainer can talk) • Material can be retained for later use • Ensures consistency of information • Promotes learner-controlled learning	• Can be boring if too lengthy • Learners read at different paces • Difficult to gauge if people are learning or even reading • Assumes audience can read or understand the language • Can lead to opportunistic breaks and side conversations
Facilitated group discussion	• Keeps learners interested and involved • Learner resources can be discovered and shared • Learning can be observed • Allows quieter types to speak up in a small group	• Learning points can be confusing or lost • A few learners may dominate the discussion • Time control is more difficult • Relies on a strong facilitator in each group
Case study	• Requires active learner involvement • Can stimulate performance required after training • Learning can be observed • Learning comes from real-world applications	• Info may not be precise or up to date • Needs a lot of time for learners to complete the work • May be affected by different levels of management/personalities in the room • Learners can become too interested in the case content • Relies on a strong facilitator
Demonstration	• Higher retention rate than lecture • Is interesting • Can give learners a model to follow	• Needs to be accurate and relevant • Examples can require a lot of preparation • Everyone may not be able to see well • Relies on a strong facilitator

1. Gain attention. Do something! Do anything to start the class with excitement. Introduce yourself with passion and show you have passion for what you're about to teach.
2. Inform learners of objectives. EHS professionals push back on this idea. Why do we have to tell them what the learning objectives are? First, because they deserve to know why they are forced to sit in the

class. (Don't worry, when you become a ninja, they will ask to go to your classes!) Second, because it helps you manage expectations. They are hearing what they will learn so they know what the class is about. I often ask learners of safety training if they know why they are in the room. Sadly, many have no clue.

3. Stimulate recall of prior learning. Take the time to tie what you are teaching to the learners' past experiences. If you know what those experiences are, you can start to connect the new concepts to the old ones. If you are forced to go in blind and don't know what their past experience is, ask them.
4. Present the content. This is when you take all the painstaking work you have done and thrill them. Try to group and sequence the content in a way the not only makes sense but will be easy to remember.
5. Provide learning guidance. Take the time to find out whether learners are getting the information correct. This is critical when you are promoting feedback and discussion by the learners.
6. Elicit performance (practice). Ask them questions about the materials and content. You can adjust the content based on feedback, and it will help you determine if the learning is sticking.
7. Provide feedback. If you want them to continue to speak and participate, then give them feedback.
8. Assess performance. Are they getting the message correctly? This is where you can determine if you and your materials are doing their job and if they will achieve the learning objectives.
9. Enhance retention and transference to the job. Ensure you equate what you teach to the job they do. If it doesn't make sense to your learners, it won't stick.

Malcolm Knowles

Knowles is considered the father of adult learning, and his six assumptions of adult learning relate more to how you need to communicate with people respectfully than to a prescription of delivery (Knowles 1990). Adults do not learn like children do. Children don't have a reference point for learning. Everything they learn they are learning for the first time; this is not the case for adults. Use the explanation of these six concepts to ensure you design your training for adult learners.

- *Why* are they at this training program? How does what they will learn impact them, what is the benefit, and what is your expectation of them for the training? Your job as the trainer is to make sure they know the answers to these questions as soon as possible.
- Your learners come into the room with life *experience*, and they want to connect what they are learning to what they already know. It is your job to help make those connections and to leverage the knowledge in the room as you train.
- Adults are *self-directed* and used to making adult decisions. You need to make sure you give them the opportunity to make decisions regarding how they learn. Not everyone learns the same way, and designing programs that allow for the different ways to learn makes room for the needed self-direction.
- Make the information *relevant* to them. Adults are open to learning about things that interest them. Making the training relevant and desirable to the audience will increase the probability that they will learn.
- Make the training *problem* centered, not topic centered. Adults are willing to put energy toward solving a problem. Making the training about a new awareness will not generate the same interest as asking experienced adults to solve a problem.
- Adults respond more to *internal* motivators, and we need to give them a training program that they want to embrace. The trainer can help with that by blocking distractions and barriers to learning.

Edward Lee Thorndike

Thorndike is another key influencer in adult education. His laws and theories of learning in the early 1900s (Thorndike 1931) are still accepted and taught. A good ninja will consider these laws when starting training-program design.

- Law of readiness: This law states that learning can only take place when your learner is ready to learn. So if the learners walk in the door not wanting to hear any more darn safety training, it's your job to change their minds and make them want to learn. You do this with good planning, using adult learning principles, and respecting them as adults.

- Law of exercise: This law is easy—The more you do something, the more likely you will remember the information. Don't get crazy with this law. Saying the same thing over and over without the rest of your learning tools won't really work since the law of readiness will block any boring repetition.
- Law of effect: This one is easy, too. Learning sticks with the learner if the experience is pleasant or satisfying, and learning is weakened if the experience is unpleasant. How you show your passion and knowledge and how you make the experience positive will result in more learning that sticks with your audience.
- Law of primacy: This law is all about what is "first," and what learners remember as "first" will stick with them. This has a downside. What if what they learned first is wrong? Ever hear that safety is common sense? People love to say that; they learned it long ago. But it's not true, is it? How could the safe entry into a confined space be considered common sense? No one is born with that knowledge.
- Law of recency: This law applies to what has been learned most recently. Have you ever asked for feedback on what learners got from one of your classes? Doesn't it often include the last major element you taught? This is why a good design that links all the elements together can help keep the information in the most recent memory.
- Law of intensity: You can guess this one. The more vivid or dramatic learning experience gets remembered. I use a small class activity in every ninja class I teach, even if it is only sixty minutes long and there are two hundred people in the room, because the impact it has on most learners is pretty dramatic. Despite the logistics of running an activity in that environment, it is worth the challenge. If you want to learn more about that activity, you will need to ask me about it!

Why Training Fails

Yes, we occasionally fail. Sometimes it's our fault, and sometimes it's not. But the only way to ensure you have done all you can to keep the learners engaged is to design and deliver the training with some of these things in mind.

They Don't Pay Attention

Did they just come off night shift? Is the material boring? Are they distracted by other work responsibilities? Whatever the cause, it's your job to overcome it and make sure they want to pay attention. Getting angry or punishing them for their lack of interest doesn't help and only further distances you from the learners. If you are not sure why no one is paying attention in a class, take the time to ask them during a break or after. Choose someone you believe can give you a fair answer and seek them out. If you ask during class, you are not likely to get the answers you need to improve.

There's Too Much Information

This is far too common a problem with safety training. We try to smash as much information onto the slides and into their heads with no regard to how much they can retain. Saying more or showing more on the screen does not increase your learners' chances of learning, and it may actually distract from the real message you are trying to send. Good design means going back to the performance or knowledge gap defined in your analysis stage and teaching them exactly what they need to know to fill that gap and achieve your business objectives—nothing more.

There's No Context for Learning

If your learners don't see how the training will affect them, they are not going to pay attention. Be sure to design in the WIIFM right up front. Explain that there has been a rash of injuries, or a policy change that will make their life easier, or that management really cares about workers and their safety.

There's Learning Interference

Phones, email, personal distractions, something interesting happening outside the window are all interfering with your ability to connect with your learners. Set up some ground rules at the beginning of class and be honest with your learners. You have worked darn hard on creating this course, and you want to be sure you are helping them get it. I usually tell my learners that I respect them, and that it is a challenge being away from the job; if it is critical to take that call or answer that email, I ask them to excuse themselves from the class.

The Information Won't Work in the Field

If we don't know our material well—how to make it relevant to the audience or how to make it work in the field—it just won't stick with the learners. All examples should make sense to them and apply to their work. Plan to find someone who can help you make sure you have done this well before you teach your course. If I know a presentation is going to contain a lot of images to help teach my content, I create a shot list with the help of an expert from the company. If you are teaching a class on the safe access to container ships from pilot boats—and you are not an expert on pilot transfer—you need to find someone who is. As soon as your audience can poke holes in the application of what you teach, you risk losing them.

What Will Your Training Program Look Like?

With your learning objectives written, the outline filling in, and your ninja brain considering how you will use adult learning principles to deliver your program, you can start to visualize what the first training class might look like. It's OK to think that way. This means that the design plan is working. Although it is not yet time to develop the materials, there are a few things to address, since you will need to consider them in the development stage.

Corporate Brands and What Those Mean to You

If you work for a company with a defined corporate brand, you will need to talk to the marketing and communications staff to get a copy of the approved presentation templates and to learn about any artistic look/feel policies that exist.

Room Setup

Do you have the space to do the training you are imagining? Is your available meeting space the right size and setup for what you want to do? Working-group activities take up a lot more space than a lecture. Is your space a space that says "great training"?

Audio/Visual

Do you have the equipment and tools to teach the class you envision, from the projection and sound system to flip charts and white boards? If you're not doing

adapt

It can be devastating to your plans if you are not ready for an unexpected change and cannot adapt quickly. Always have backup activities and be ready to rearrange your working groups to accommodate such a moment. Overplan field trips outside the classroom, and keep the disruptions under control and safe. And be sure the work you want to evaluate is being done that day.

a monotone lecture anymore, you will need the budget, space, and expertise to deliver the newly designed program with excitement and enthusiasm.

Field Training

Some of the best training is the kind where learners get into the problem-solving of safety right on the plant floor or construction site. The aha moments can be profound, especially when workers can solve a safety problem they never saw as a problem before you taught them how to see it. But as valuable as it is to leave the classroom, it is critical to plan these activities. You must coordinate with the right people to ensure you do not disrupt workflow and that you have the right support from the right people.

Assets and Resources

We all need more time, both to develop better training and to deliver it. Unless you have the actual time length of the course specified by a legal or regulatory requirement, you will have to determine how much time it will take to deliver the training properly. Let's assume in a perfect world you get the whole day to deliver the orientation training for all new employees. In that eight-hour day, you don't actually get eight full hours of instruction time, taking into account breaks, lunch, and other noninstruction tasks. These need to be considered during design.

Some instructional designers will have the luxury of dictating how long a course will take to deliver based on the design roadmap, using the business and learning objectives. Essentially the course is as long as their outline of the program says it is.

Although that would be a great way for us all to work, it is not the typical experience for the EHS professional. So, as you design, consider the time it will take for the following:

- Introduction: If you don't know your audience, you will need to go through introductions for the benefit of learners and yourself (think mini-assessment on the spot as they introduce themselves).
- Administration side of the course: This is the learning objectives, emergency procedures, review of learning materials, and the "rules of engagement" for the day.
- Icebreakers: This could be five minutes or much longer. One of my favorite icebreakers usually takes more than thirty minutes for a class size of twenty-five people broken into five working groups. That is a big chunk of time, but it's a critical element of my program for some types of training.
- Breaks: Are they fifteen or twenty minutes, and how many do they get?
- Lunch: Thirty minutes or sixty? Do you schedule lunch for sixty minutes and then offer to finish early if they take only thirty? Who doesn't like to finish safety training early?
- Activities: This part is huge, and we will discuss it more in the development section.

I use a spreadsheet to calculate the teaching time available. In figure 4.1, you can see the start time of 8:30. First, I fill in the times for lunch and the breaks and then estimate the opening and closing sections, including time to do evaluations. Once those are complete, I know what I have left to work with. In this sample table you can actually put in the time you think each slide should take, and it will autofill the schedule as you add, subtract, and edit your program. Your goal is to know how long you plan to spend on each slide of instruction and each activity so they will end at roughly the time you planned. I usually add some cushion time when determining how long an activity will take. When you break down your course to this level of detail, you will see just how tight your day is. I typically do a longer, single break in the morning if the clients support that idea, followed by two shorter breaks in the afternoon, since it's much harder to stay awake later in the day. I geeked out on this example, even color-coding what type of activity I would do. The color distribution helps me see the bigger picture. You may notice that the training ends at 4:11 PM. I usually give myself a buffer of about fifteen minutes to allow for extra discussion among learners or if I didn't plan for enough feedback time during group activity report outs. If I actually end early, that's OK; no one is ever unhappy when safety training ends early as long as you meet all your learning objectives. You can find a free training time calculator to download and a short video on how to use it to your best advantage

Slide	Topic	Start	Minutes	End
1, 2	Welcome/Emergency directions	**8:30 AM**	5	8:35 AM
3	Introductions	**8:35 AM**	20	8:55 AM
4, 5	Video - Embrace the safety	**8:55 AM**	2	8:57 AM
6	Learning objectives	**8:57 AM**	3	9:00 AM
7	How it works	**9:00 AM**	3	9:03 AM
8	Video - Ball game	**9:03 AM**	12	9:15 AM
9, 10	Activity - ID safety challenges	**9:15 AM**	25	9:40 AM
11, 12	Video - Change	**9:40 AM**	1	9:41 AM
13	Learning - Bradley curve and interdependence	**9:41 AM**	10	9:51 AM
14	Activity - Change for the good	**9:51 AM**	25	10:16 AM
	Break	**10:16 AM**	20	10:36 AM
15–17	Learning - Communication	**10:36 AM**	10	10:46 AM
18	Activity - Karma job aid	**10:46 AM**	25	11:11 AM
19, 20	Learning - Power types and how to influence	**11:11 AM**	10	11:21 AM
20–22	Learning - Influencing skills	**11:21 AM**	10	11:31 AM
23	Activity - Influencing job aid, mini	**11:31 AM**	10	11:41 AM
24–33	Learning - Influencing skills, full	**11:41 AM**	60	12:41 PM
	Lunch	**12:41 PM**	30	1:11 PM
34	Activity - Influencing job aid, finish	**1:11 PM**	20	1:31 PM
35–39	Learning - Why change?	**1:31 PM**	10	1:41 PM
40	Video - Dancing man	**1:41 PM**	5	1:46 PM
	Break	**1:46 PM**	15	2:01 PM
41	Learning - Your safety team	**2:01 PM**	10	2:11 PM
42	Activity - Changing the culture	**2:11 PM**	25	2:36 PM
43	Learning - Bradley reminder	**2:36 PM**	5	2:41 PM
44–46	Learning - Goals of caring and you	**2:41 PM**	10	2:51 PM
	Break	**2:51 PM**	15	3:06 PM
47	Activity prep	**3:06 PM**	5	3:11 PM
48	Choose a change agent buddy	**3:11 PM**	10	3:21 PM
49	Activity - Complete traffic light	**3:21 PM**	10	3:31 PM
50	Evaluation	**3:31 PM**	10	3:41 PM
51	Activity - Read traffic light and discuss commitment	**3:41 PM**	30	4:11 PM

FIGURE 4.1 Course time estimator

FIGURE 4.2 **Sample QR code for The Learning Factory channel and the link to the training time calculator**

at https://www.youtube.com/channel/UCOLL3mDmBY5kqHZxUoeiI3Q or by scanning the QR code in figure 4.2.

Other Asset Management

Other elements of the program design need to be considered early in this phase as well. These items are critical but sometimes get forgotten until later.

- Materials: What job aids, handouts, manuals, books, regulations, or standards will you be distributing? Will you make the copies in-house or do you need to send material to a printer? Will you need to purchase anything in advance to distribute, such as ANSI or ISO standards?
- Tools: If you will be training on a new process, tool, or piece of equipment, will you have all the components needed to teach effectively? If anything is used in the field or on the shop floor, how do you plan to demonstrate the use properly? Regarding training tools, does your company have the meeting space, screen, projector, and any other equipment you will need to deliver the training? If you are teaching in the field using technology, will you have the electricity and space to ensure you meet your objectives?
- Vendor-provided props: Sometimes vendors will come in and help deliver the training or simply provide sample equipment or PPE. Be sure you have gathered any donated materials early so you are not scrambling at the last minute.
- Food and coffee: You need plenty of coffee, no matter where you train. Is it provided automatically in a breakroom? Do you need to brew

it yourself, or will you have help? Will someone make sure there are refills before the scheduled break? Do you need to contact a caterer and have it delivered, or will you stop by a coffeeshop to bring it in yourself? These seem like little details, but I have seen programs get off to a very rocky start when there is no coffee, and the training suffers because of the learner distraction. Food and safety training go together. Find out what the menu will be, and make sure the delivery time matches your planned lunch break. Encourage whoever is managing the food to find out about potential food allergies or religious requirements of learners. It may not seem like this should be your problem, but everything that happens during the training is tied to you and your course, and unhappy learners—or worse, hungry learners—are a tough audience. If no food will be provided, let your learners know in advance so they can make plans to either bring their lunch, or provide some time to leave the class to eat elsewhere. Eating lunch offsite can make getting learners back on time difficult, so be sure you have scheduled enough time for this, and, if you can, provide a list of close restaurants to get them back to class quickly.

Training Program Development Outline

Z490.1 recommends you develop and maintain a course outline as part of the accepted practices in EHS training Annex B.14 (ANSI/ASSP Z490.1, 36–37). Not every component will apply to you, but it is a good idea to start thinking about each element. Consider why and how you need to track, as well as why you develop the training you do. Create the outline during the design phase of the training plan, but be sure to remember to update the records as the program changes and improves. Elements to include are listed in table 4.5.

Designing to Shine

Are you ready to train like a ninja? Do you have your skills and confidence ready to be the ninja you are meant to be? In the past, most safety training was viewed as a necessary evil, and many learners and trainers still view it that way. The only way your learners will recognize things are different now is when you tell them and show them. The "show" part is coming up in

TABLE 4.5 Z490.1 Program Development Outline*

Data to Include	Description
Course title and identification	This includes any ID numbers you need to reference for your LMS or SMS.
Publication date	This is the date the materials were finalized.
Scheduled course length in weeks, days, or hours	For the EHS world, I think it is best to track in hours, since some governmental compliance training and other ANSI standards dictate hours.
Purpose and any regulatory drivers	This can be valuable for your own SMS records, but may also prove helpful if a governmental, contractual, or auditing body wishes to see your records.
Overall learning objectives in terms of anticipated behavior, working conditions, and performance	This one should be easy, but maybe you see the theme that good LOs are important to good training.
List of course attendance prerequisites	Not only good to write down, but important to keep track of so no one's time is wasted when they go to the wrong training or training they are not prepared for.
Training locations	Be as specific as you can, list addresses when possible.
Trainer requirements (training aids and operating equipment)	What types of equipment or aids will you need to deliver the course effectively? I have a standard list that I add to depending on the course.
Equipment requirements (training aids and operating equipment)	If this is a field-based or hands-on training program, you will need to identify and list all the equipment you will need to be successful.
Space requirements (by type, capacity, and number)	It's not just a room anymore. Do you need break-out space for working groups? Will you set up your space differently than you did before you became a ninja?
List of learning objectives in terms of duties, tasks, and job elements	Make sure you can actually achieve the learning objectives in the class; do you have all the equipment, tools, job aids you need to succeed?
List of required reference materials	Do you reference government regulations, operator manuals, resource texts, or anything you should have with you when you teach?
List of evaluation instruments	What kind of system will you use to determine if the training is effective?
Sequence of instruction by trainer, guide, title, and number	This is designed to track when you need to keep the data in your LMS or SMS. This unique identifier helps keep track of multi-element training products. Don't worry, if you have these "unique identifiers," you know what I mean here.

*Adapted from ANSI/ASSP Z490.1-2016, 36–37.

chapter 5, but you should start telling them right away about the new training that is coming.

Advertise Yourself and Your Skills, Not the Program

Take the time to share the information about what you are doing differently in the design, development, and implementation of future safety training. Send an email to the people who need to know, and explain all the changes you are making to the safety training. Think about which stakeholders will be helpful and will support the new training style, and get them on your team. Are there any influential workers or union representatives who act as leaders who can help you promote the new way to train?

Advertise the Program with "What's in It for Me?" (WIIFM)

Promote what you are doing differently, not just that the program is different overall. Be specific and share the upcoming dates of the new training. Include information on the learning objectives, who and why someone should attend, and, of course, promote any of the fun and exciting ways this program will be different from past training.

Remember your adult learning principles? Get them excited about the program any way you can. If they see a value in the program, an opportunity to do their job better and safer, and that the program is going to be great, you have set the stage for your success.

If we, the trainers, treat the program as a last-minute, don't-have-time-to-do-it-well training program, there is no way our learners will treat our content as critical to their work. So, starting now, this is the *real* training. You are a ninja, and your course will be not just different but great.

As we close on design, you may notice a lot of holes still in your safety-training plan. That is because the next chapter is about filling in the blanks completely. Design is your roadmap, your new ideas, the plan to do things differently. Now it's time to fill in those blanks!

References

ANSI/ASSP Z490.1-2016. 2016. *Criteria for Accepted Practices in Safety, Health and Environmental Training*. Park Ridge, IL: American Society of Safety Professionals.

Bloom, Benjamin S. 1956. *Taxonomy of Educational Objectives, Book 1: Cognitive Domain*. New York: David McKay.

Gagne, Robert, and Karen L. Medsker. 1995. *The Conditions of Learning: Training Applications*. Alexandria, VA: American Society for Training & Development.

Knowles, Malcolm. 1990. *The Adult Learner: A Neglected Species*. Houston, TX: Gulf Publishing Company.

Thorndike, Edward L. 1931. *Human Learning*. New York: The Century Company.

Chapter 5
Develop Your Training Program by Filling in the Blanks

Now it's time to open PowerPoint.

Take those carefully crafted learning objectives and put them right on the first page. Copy each one separately and place it on a new page. Now it's time to fill in the blanks. You can clean up the wording later. Those learning objectives are where you begin to think about content. They can also serve as your "test, then teach" script. As you comb through all your content sources, you may find something you think is good to include but doesn't fit under any of your learning objectives. Guess what? It's time to determine if you really *need* that content or if it's just something nice to have. Ask if the learners need it to fix the gap? If the answer is no, it doesn't go into the materials.

Gather all those notes, reference materials, and thoughts that have been bouncing around in your head and start creating content. But be careful! Don't fall back into the trap of using existing PowerPoint slides from the internet or your company files. If those files have great material worthy of appropriation (legally), then only select the *parts* of those other materials that support the learning objectives to include in your content.

Where else should you look for content? Here are some options:

- your own knowledge
- results from your training analysis
- your trade or professional association
- information from the learners and their stakeholders
- shared content from ASSP's practice specialties and common interest groups

- government regulations and supporting training materials
- consensus standards, such as those approved by ISO, ANSI, OSHA (or whatever exists in your part of the world)
- company policies and procedures
- videos you create showing how to do the job the right way
- purchased content that you can use and edit legally
- accident investigation and near miss evaluation results
- job safety analysis content
- equipment training manuals
- vendor-provided materials
- equipment-usage job aids

If during the collection of all the technical content you find a weakness in your design, good. That means ADDIE is working! Take the time to go back and edit what you need to, improve your learning objectives, or further define who the learners will be. We have already covered a long list of things to consider about your learners. Don't forget to review what you determined about your audience so you are developing the materials to match who they are. This is also the time to review your evaluation plan. Will what you are developing achieve the learning objectives? Does the evaluation plan still work with the content? Remember that ADDIE is a cycle—it can handle any adjustment during any part of the cycle, even after you teach the class.

Plan to Be Your Best Ninja

Icebreakers

Icebreakers can be very useful tools to help begin the process of learning, but they can also be a waste of time. I have participated in icebreakers that annoyed the learners because they had no relevance to bridging the learning gap. If you're unfamiliar with icebreakers or haven't tried to use one, it is simply a way to set the tone and expectations for a class by getting everyone involved in learning right away. Icebreakers can serve many purposes, including getting to know the learners if you don't already.

Icebreakers also can:

- act as the first attention grabber of the class
- let your learners know you want them to participate in their learning

- get the pace of the class going (this won't be a boring lecture!)
- help you and your learners feel at ease
- help the other learners get to know each other better
- get everyone to participate, even the quiet learners
- help to set your class culture (more on this later)
- start moving the training toward the technical content

One icebreaker I use for soft skills and leadership training involves drawing on flip charts. As teams, the learners have to design and draw familiar objects. The difference here is that the last object they draw is imaginary—a word that I make up. The teams are forced to join together to create something that does not yet exist. I provide no help with their drawings other than naming the objects. Because the last object is not yet real (until they make it real), the teams must work together and think outside of the box. The result is lots of laughter and team building. Then I close with a question to the learners: Are you ready to learn new things, to be open, to think outside the box and try learning in a new way? Almost everyone always says yes! And that propels me to start my new training class, with open minds and laughter.

Managing Expectations

If you are launching your first ADDIE-based program, you should let your learners know they may be in for a change. If you are going from lecture-based, word-filled PowerPoint slides to a newer ninja safety training program, you will want to let them know about the change. The first time you tell them will be when you advertise your program (yes, you should, and we will talk about that soon). Tell them again when you begin your training. Even though the new program will be awesome, it will still come as a surprise to some, and invariably there will be a person or two who won't like the new way. Remember that you can't please everyone, but if you are pleasing more than you used to and they are learning the materials, then you are on the right track!

During my personal introduction, I always let the learners know about my teaching style (passionate, fun, interactive) and how the class may be different from what they have experienced in the past. When I ask them to introduce themselves, I might ask what they expect from the class. I encourage them to be honest; if they have not appreciated EHS training in the past, I see this as a challenge to open them to a world of effective, engaging training. If you don't have the time to do class introductions, or you are not sure they are ready to be that honest, you can ask some questions about past experiences, and they can answer with a show of hands. This ninja has found that facing any negative expectations or learning obstacles right up front is the best way to get learners to open their minds to a new type of EHS training.

Test, Then Teach: A Ninja System to Keep Learners Engaged

After working several years with the same client, we developed a system to keep future training relevant and fast paced, according to each class's needs. During the development stage, we begin each new learning section with a working-group activity. We ask teams to go to their flip charts and answer a question that relates to our learning objectives. For example, in a confined space entry refresher course, I would ask each team to list five steps of a safe entry indicated by their company policy. I might reward the team that answers the fastest or the one that gives the most accurate answer. Honestly though, I usually reward everyone with praise and maybe a treat for participating, especially if this is a new concept. This method allows me to "test, then teach." This is successful for several reasons:

- I can adjust what I teach based on what they know about the material, since they have just shown me what they know by completing the working-group activity successfully. If they cannot complete the exercise, then I can launch into a mini-lecture about the content. Better yet, I can provide the policy and have the teams fill in what they missed on the flip charts. Either way, I go from a straight lecture on material they may know and don't want to be lectured about to a fast-paced, learner-led activity that identifies exactly what they know and don't know. Since a ninja always comes prepared, be sure you have all the materials ready to teach if they still need to learn the details and have devised a smooth way to skip the materials if you don't need to teach them. I usually say something positive about how smart they are and that we now get to jump ahead since they have shown that they know the material.
- This method allows the teams to teach each other or learn from the provided company policy, instead of the trainer lecturing on slide after slide. Some people might cringe at the thought of having learners read policies, but you can limit the time and shape the goal of the activity so it doesn't exceed more than a few minutes of reading. I

think that listening to a lecture all day is pretty cringe-worthy; using this activity at least lets you try different adult learning principles.

- You become a facilitator instead of a lecturer—always a better way to help learners.
- You can watch the groups as they complete the activity and start to see who knows most about the topic, who are the natural leaders in the team, and who you may need to ensure gets extra help. This fourth benefit applies to any activity you conduct, so make sure you engage with the teams and offer help, guidance, or instruction, as needed.
- You can confirm you have met all or part of a learning objective, and you have it on paper. You could even take a picture of the completed flip charts for your training records, with the learners' names on the sheet, or include the learners in the picture!

Teach, Then Test

The alternative to the above method is more common but may result in a default lecture. If the information is completely new to the learner and you must provide the information in a lecture format, then teaching first may be your best choice. If your subject matter is a new government regulation, technical corporate policy, or complex machinery, teaching the materials first may be the only way to get the base knowledge to them. Remember, just because it's new doesn't mean it has to be lectured. Get creative and use some of the other methods I discuss below.

Z490.1 says to be sure to develop your program based on the following:

- Introduction - Present the overall picture. Be brief and focus on specific critical training objectives. Let the audience know how they will benefit from the training and what will be expected at the course completion. Tell course participants why they are being trained.
- Main body - Present the required and desired information. This is where the majority of information is given. All regulatory, safe practices, and best business management practices should be given during this portion of training. This is a good time to apply useful transitions and memory joggers. It is recommended not to wait until after lunch to begin this portion of training. Movies, lectures, and sit-in-place activities should be avoided directly after participants

have eaten a heavy meal. Hands-on and motor skills training often work best directly after eating.

- Conclusion - The conclusion should be planned and rehearsed. An interested audience will usually remember a high-impact closing statement. Remember that safety, health and environmental training technical content is important, but it alone will not keep the interest of the audience. The final impression should be a lasting impression. Always try to restate the training objectives during closing statements. To make training more memorable, experienced trainers often use quiz games, hands-on scenarios, and other group activities prior to closing remarks. (ANSI/ASSP Z490.1-2016, 42)

Methods of Delivery

So far, everything in our design is pretty inexpensive to do, but we are now getting to the design stage where money may become an issue. Better training doesn't have to cost more, but it could. Bad training can cost a lot of money too, either in wasted resources because no one learned anything, or poor design that didn't leverage the money available. Many EHS professionals don't know a different way to train other than by lecture. It's how they were taught, everyone they know teaches that way, it's taught that way in college—so it must be the only way to teach safety, right? *No!* There are so many other great choices for teaching your team what they need to know; they just take some planning. If you choose to use something other than a lecture, however, your preparation must be thorough and detailed. This is just another step in your program development.

Lectures

You knew we had to talk about it eventually: the dreaded lecture. Lectures are the reason most safety training is ineffective and, at their worst, are the reason people do things unsafely. If no one listens to the lecture because it's boring, irrelevant, or unneeded, we lose our chance to influence the behavior or to take advantage of a knowledge-change opportunity. We also guarantee that most learners will not look forward to future safety training.

The great news is that there are many other ways to improve safety through training that do not involve a lecture, or at least not a bad one. What is critical in the design stage is deciding what elements you will use to develop and deliver your materials. This is when you go back to your learning objectives and ask yourself "How could I deliver this in a way that will make it easy and interesting for the audience to learn, even if it is a regulatory or legal topic?"

Lecture-bashing aside, lectures do have their place in EHS training, including:

- for a fast, just-the-facts type of presentation
- as part of a bigger training program with many alternatives for using ALP
- for keeping control of the class due to the complexity or seriousness of the materials
- when a guest speaker or stakeholder has a specific message to share
- when your audience is so big you will not be able to encourage group activities
- if it's a really great lecture (hey, it can happen!)

I like the way my friend and book mentor, Elaine Biech, calls them "lecturettes" in her book *Training and Development for Dummies*. Her made-up word "gives the illusion of being less tedious and a bit more playful" (Biech 2015).

Alternatives to Lecture or Lecturette

There are many ways to use an alternative to the lecture. All the alternatives require more up-front work during design and development, but this work will allow the facilitator to relax a bit during the activity. It does require that you carefully plan the timing of the activity, from providing directions and completing activities to getting group feedback. It's a very satisfying moment for a ninja when all the teams demand a chance to provide feedback and

reminders

Think about the worst EHS training you have ever attended. Take five minutes and write down exactly what made it so terrible. Keep that list on the wall near your desk. Look at it often, and remind yourself not to do those things. Simple reminders can help stay out of the boring lecture trap.

share what they've learned. Be ready to adjust your timings and content if this wrecks your schedule—it's probably worth it.

Look at the options below and some of the examples. Could you incorporate some of these hands-on adult learning options into your program?

Demonstrations

We know that doing things that apply to the job and training materials allows for greater learning retention. In most cases, this gets learners up on their feet and allows practice of a new skill or application of knowledge. Keep in mind that some of these activities involve going into the field or onto the floor. This can take a lot of time but will be worth it. Also consider that if you take your learners into an active work area, it may affect the production of the work in progress. Make sure everyone affected knows your plan and supports what you are doing. Be sure that taking a group of people on the floor or into the field does not create any hazards; sometimes a group of eager learners can be distracting to others doing their jobs. Also, ensure the work you want to show is actually being done at the time of the activity. Nothing is worse than finding out the task you want to watch isn't being done that day or it's break time with no work being done. If you will be doing demonstrations and teaching future classes on the same topic, you can use the demonstration activity to generate interest within your class for future learners.

Here are some different types of demonstrations:

- Hands-on. Bring out the props! Use any tool or piece of equipment to teach your learning objectives. Choose anything from safety glasses to hearing protection, SDSs to the actual chemicals (choose this only when safe), fall protection harnesses to tools or equipment. Even vendor miniatures of products work well. This list is endless, depending on the type of work your learners do.
- Shadowing. Have the learners watch someone do the work safely first if it is complex or can only be shown on the floor or in the field. They can then watch the skill in a safe manner and, if the situation allows, they can practice the skill.
- Field trips. Do you have a sister company or site where you can teach a skill to the learners in a new setting? Sometimes seeing a hazard and the solution in another environment opens learners' eyes to what they miss at their usual place of work.

FIGURE 5.1 Sample QR code

- Computer demonstration. Some of the learning we will do involves software. If you can show it in real time and not with a bunch of screenshots, the learners can practice the skill to proficiency. Screen shots are hard to memorize, but if you must use that method, you may need to develop a job aid for the process that they can use later as they work.
- Video demonstration. Smartphones today can produce some pretty great short films. Don't worry about great audio; you can narrate over the film in the class. Film a qualified person doing the job in a safe, productive way, and show it to the learners. Slow it down or stop the film at critical steps, and promote discussion on safe task completion. Make the film available as a job aid later in case they need it. Store the video on your intranet or other IT-approved location, then create a QR code sticker like the one in figure 5.1, and place it where the workers can use their smartphones to see the video any time. Have you ever seen one of those weird square boxes that looks like it's full of static? That's a QR code! It's sort of like a barcode, and it can hold almost any text, links, or information you want.

Activities

Some activities can be completed in the classroom, while others can be done right in the workplace. Keep in mind some of the special considerations of training near live work (mentioned above). Through the years, many budding ninjas have struggled with trying bold new activities, so this extra information may help you.

Problem-solving

Describe a likely hazard learners could experience and show them how to use a policy, procedure, or governmental regulation to correct the hazard. It's great to invite nonexperts on the topic into the activity. An outsider's perspective can be astonishing for solving problems that are too close for the learners to see.

Case study

Use a near miss example to challenge learners to determine ways of preventing other near misses or actual incidents. You could also use information gathered from audits, walk-arounds, safety inspections, or items from safety suggestion boxes.

Hazard hunt/fix

This type of activity can be incredibly successful if planned correctly—and it requires a lot of planning. Send your teams onto the factory floor, out to the shop, onto the road, or into the dirt. Have them find and design solutions to hazards they identify, as they relate to your training materials.

These activities can be time-consuming and disruptive to others but can yield amazing results. Work with your stakeholders so they know what to expect. Production slowdowns and disruptions are just the beginning. Stakeholders need to be ready to respond to the hazards and solutions identified. Nothing kills your safety efforts faster than having the higher-ups ignore what the learners came up with in class. Don't forget to share the problems or solutions as they occur going forward. This reinforces the training and validates the importance of learner participation.

I try to send learners to areas that are not part of their work area or task so they are not "blind" to the familiar. If the learners see something that must be fixed on the spot, be prepared to handle that quickly. You can't have them identify a life risk, then walk away and talk about it later in class. If your class is the kind that can offer interventions to others as they see them, encourage them to use the training time to fix what they see. I have had learners come back and report out that they noticed other safety hazards outside of the class scope. For example, a learner noticed someone was not wearing proper PPE, and when asked what they did about it, they looked at me blankly. Lucky for me, a fellow learner interjected, "We'll go take him a pair of safety glasses!"

Critical incident evaluations

This can require more working groups in the classroom but may involve visiting the location of the critical incident. Be sure you have all the facts to allow learners to conduct a full evaluation. For more complex incidents, you may want to put together a package for the teams with all the relevant information, including facts, pictures, and even props (such as the fall protection harness in the earlier example or a copy of a company policy or procedure). Don't put all the materials on PowerPoint. It will likely be too hard to read and several slides long. Consider if you want to use different examples for each team, or the same example across all teams for comparison. Both are good choices, but different examples require much more report-out time.

Brainstorming

Offer up a hazard that has been tough to fix or a new part of a regulation that must be implemented. Have the teams come up with solutions and share them with the class.

Role play/coaching

This activity may require some serious encouragement on your part as the facilitator. Not everyone likes to get up in front of the room and role play, while others can't wait. When I use role play, I either pair up teams and ask for volunteers to share what they learned in their pairs, or I have larger teams work as a group and nominate who will deliver the role play to the class. This allows the less outgoing learners to avoid embarrassment. I also wait to do this activity until after I have learned more about my learners via an icebreaker *and* other activities, when I know who might be comfortable in front of the rest of the class. Role play and coaching are excellent for teaching soft skills like praising for safety, effective interventions, and EHS leadership skills. I cannot think of a time in my career when a role-play activity did not lead to enormous amounts of fun while learning. It's a great way to break through some of the serious content we often teach. It is also great to do after lunch when even the best ninja struggles to keep the class engaged.

Recently I was doing some role-playing on the reading of body language and its value in safety coaching in the United States. It turned out that we had five pairs of learners who spoke other languages. They completed the role-playing while speaking their native language, not English. This allowed the other learners to focus only on the body language, not the words used.

What an experience! The role-players had great fun, and the other learners were able to "see" the conversations taking place strictly via body language. Six months after the training, I was still receiving emails from the client on how valuable the experience was.

Game-show adaptation

This is useful if you want to measure learning during the delivery of your program. A game-show adaptation is much more exciting than a test! Name that hazard, find the hazards in a picture, or a jeopardy-style competition is a great energizer for the class. Do an internet search for the type of game you want. You can find examples of premade PowerPoints that only require you to insert your questions and the answers. This is also a great activity to do after lunch to review.

Teach back

After you have shared new materials with the class, have them develop a mock training program to teach back to the class. For example, you are teaching your class about the new walking surfaces regulations. Following your "lecturette" on the materials, provide the regulation and assign each team a different part. Have them teach the class what they have learned and challenge them to find something you missed when you taught.

In 2014, I was facilitating a safety supervisor class in Malaysia. Although everyone was an English speaker, it was not the first language of anyone in the room, except for me. After two intense days of facilitating, we got to the teach-back activity. Each team had to demonstrate how they would be delivering the materials back to their learners. One team was from Thailand and worked so hard to understand and absorb all the knowledge they could. They were teaching back a section on eye protection. I acknowledged their hard work and diligence in the class, then asked that they teach back the class in Thai. I don't speak Thai, and neither did anyone else in the room, but there was no doubt what that team was teaching us. They were passionate, excited, animated, and most important, comfortable, because they were speaking their own language. I may not have done the same thing if the teach back was on the proper shutdown procedures of a nuclear power plant, but the rest of the class loved the group's presentation, and it put all the learners at ease during the big final activity. It was a great reminder that it's all about the learners—not the facilitator!

Commitment statement

Many companies like to use commitment statements as part of the learning and learning measurement. One client likes the traffic-light approach. You might design a fill-in-the-blank handout that states:

I will start (green light) doing this _______________ as a result of this training.
I will improve (yellow light) this_______________ as a result of this training.
I will stop (red light) ________________________ as a result of this training.

What you do with the commitments after class can be useful. Do you keep a copy and go back in a few weeks and check in with the learners on their progress? Do you report that progress to the stakeholders? Maybe you do if that was a part of your design plan. Tell everyone about that measurement plan up front so it's not a surprise to them later. Some companies I work with make the commitment statements part of their performance review. Others ask the learners to post them near their workspace and to encourage others to keep the improvement challenge going after the class. Some classes get back together after a few weeks and share how well they are doing on fixing their gaps. What other ways could you use a safety commitment page in a positive way to measure and reinforce learning?

What have I learned?

This is a favorite of mine because it incorporates so many elements of ALP: energy, competition, real work application, law of recency, stimulation of learning recall, allowance for feedback to the facilitator, and so many more. It's a simple activity. After a break or lunch, or the next morning, ask all learners to write one thing they have learned on the flip chart. It must be something they just learned during the training (so they can't write down something like "safety is good"), and no one can repeat what anyone else has written. Then I have each person tell me what and why they wrote what they did.

If you try this, there are a couple of things to keep in mind. Don't be offended if they start looking in their handouts to find something to write down; even the best trainers have to accept this. Maybe they will retain what they write down. And be careful of the mad rush to the flip chart. Those who listened to your directions will realize the last person at the flip chart has to dig deep for an answer. When I lead the discussion of what is on the flip chart, if one item looks very similar to another, I let the class decide if they will accept the answer from the other learner. Sometimes they vote it down and make them write another. Occasionally, the learner can talk his way into crowd acceptance. Imagine that! A learner in an EHS training program arguing to keep what they wrote down by proving that they learned something. You can smile your secret ninja smile at that moment because you are doing what you set out to do.

Self-Study

Readings

I'm not talking about a lot of reading or assigning the reading of a new government regulation. No one is going to do that, so why assign it? Think in terms of chunks of information that they can read, learn, and then discuss. It's a nice break for you, and it lets those learners who learn best by reading utilize their favorite technique.

Crossword puzzles

You can go online and use free software to create nice crossword puzzles. They can be a great way to pretest knowledge about a topic without calling it a test. I used one once when the learners were assigned to do a huge amount of reading on their way to a conference in Las Vegas. I knew no one was going to read all that material, especially on the way to Vegas. They needed to know the materials for the course to be successful, but I didn't have the time to teach everything they should have known before arriving. I designed a medium-difficulty crossword puzzle and let each table work as a team to find the answers in the materials. It took about forty-five minutes to complete the activity. They probably learned as much as reading the material on a plane would have provided, and it was a great icebreaker for a room of several hundred people.

Pretests

Sometimes you have to give a pretest. Your stakeholders say so, the law requires it, or you want it to prove you taught them something. I would recommend using a different word than *test*. It sounds like you are back in school, but these are adults. Don't try to trick them on the test, and don't include things that have nothing to do with what you are going to teach. If you plan to give the same test after the program, be sure you designed the course to teach everything on the test. Regarding grading and retaining test results: Be sure you go over the answers on the final test and have everyone review their tests and correct any wrong answer to 100 percent. Don't just allow them to change the answers. The learners should actually relearn the information as you teach it to them again during the test correction phase, so as they correct the test, they really will know the correct answer. Legally, it looks pretty bad if you have wrong answers in the files and someone gets hurt because of that exact incorrect answer later.

Self-analysis

I use this option more with EHS soft skills. Having your learners reflect on what they already do or don't do on paper can be valuable. Questions such as "How can I improve my intervention techniques?" or "What ways do I set a great example for safety?" can spark some self-improvement opportunities. I generally let them keep that information private and use the paperwork as they think best.

Creating Your Activities

As I have mentioned, creating great activities is harder than it looks, especially when you first start. If it is going to be an activity where the learners need a lot of information, like a case study or risk improvement strategy, always create a worksheet for that activity. Write down all the directions clearly, include any images and information they need, and then leave space to complete the activity. Also include how long they have to complete the activity. Next, ask someone you trust to review the information, and ask that person to complete the task. They need to try to complete it because, if they can't, you must fix what is wrong. You don't want to learn you are missing information or have unclear directions in the middle of a class. Finally, find someone else, a potential learner or maybe someone who helped you in the analysis stage, to review it as well. They can help make sure it will work for the audience. One ninja rule I always share is: whatever time you think an activity will take—double it. If it involves a field trip to the shop floor or to the dirt—triple the time. A great activity with a poor time estimate can kill your schedule. This makes you rush later or skip material, and what if you really need to cover that material? You can't ask the class to stay late because you misjudged your timing. But no one is ever angry when EHS training finishes early!

I've covered a lot of choices for different activities. I hope several sound interesting to you and will fit into an upcoming project. But how do you choose which activity is right for your class?

- Does it have purpose and value? Make sure it is more than simply fun. Make sure the learners can look back at the activity and state what they have learned, even if you don't require them to do that.
- Does it support your learning objectives? This is when you look at your outline and ask if you need to do this activity to teach the learners what they need to know.
- Will it fit into your schedule? We talked about building your schedule during the design phase. After you rough out an idea for an activity, determine if the time it will take fits into your schedule, or if you need to choose another that does.
- Does it support adult learning? Look back in the last chapter. Is your activity supported by ALP? If you are doing several activities, try to incorporate different aspects of ALP into your development.
- Can everyone participate effectively? Are there language or reading barriers that may impact your plans? Did you remember to tell everyone to wear clothing and PPE that matches an activity you have scheduled? Also, consider if you have executives or other management in the room whose presence might discourage open participation. If they are supportive of your work, they can also serve as great examples of how to participate.
- Is it right for the class size? Do you have too many people to allow for the feedback and interaction needed to do the activity? Do you have enough people?
- Is it right for the classroom? Having the class leave their seats and move around the room is great, but what if space is so tight that this becomes a safety issue? If your room is too small, plan to borrow extra available rooms just for the activities. Even a break room or trailer can do the trick. But remember, you will need to remind them to be back in the classroom on time.
- Does it apply to the real world? There is no point facilitating an activity that they can't use on the job. Do your best to draw a connection, even if the hazard hunt they do is on a piece of equipment they won't ever use. This doesn't mean the process of finding and fixing

hazards is the same for everyone, but don't make them shave for a respirator they will never wear.

- Can you adjust the activity on the fly if it starts to fail? Practice, practice, practice your activities. Find someone who really thinks that activities are a waste of time and have them punch holes in your plan. Maybe they are right, but if not, you are a stronger facilitator after they try. If it starts to run too long, what is your plan to speed it up? If the activity is running long, I ask the remaining teams who are reporting out to pick the one thing they want to share with the other groups. If each team is doing the same activity, I remind everyone only to report on new items not yet discussed by other groups.
- Can you write directions for the activity so it can be completed without having to ask for help? No matter how many activities I have created, I always find ways to improve my directions (which I do for the next class). For more tricky activities, I actually do a mini-example before I break them into their teams. This often clears up any confusion.

Choosing the Teams for Activities

There are several ways you can divide your learners into teams. You will have to plan for what is best for the topic and your audience.

- Random. If your class has a nice mix of personalities and experience, you might just let them count off to create the number of teams you want.
- Deck-stacking. Sometimes you have to plan your teams in advance and assign learners to those teams. This could be done to keep the jokesters in separate groups, to keep line staff on a different team than their manager, or to disperse different levels of experience so each group can do their best possible work. In some of my leadership classes, I need to create the right mix of technical and soft-skills people to maximize learning.
- Let them choose. This usually leads to chaos, so I don't use this method or recommend it.

If I am doing a good job designing the course for maximum learning, I might need different-sized teams for different activities. I would not change the team sizes more than once in a class because it confuses people and makes it hard for them to trust the program. For example, for a discussion on feedback intervention, I may instruct them to pair off with the person next to them starting on the left side of the room. Later, I may use preassigned teams for the hazard hunt and fix. But I wouldn't make any other teams for that class.

Designing Your Room to Make Activities Work Best

Many of you will have to work with what you have, that is just the way it goes in the EHS training world. But even if your choices are limited, you should know about what options are available and try to incorporate as much room design as you can. Table 5.1 shows several different kinds of classroom layouts and lists some pro and cons for each.

TABLE 5.1 Room Design Options

Name	Example	Pros	Cons	Comments
Theater	*theater style*	Ideal when teaching to large groups, for short lectures, or when a screen is required for the duration of the course.	Not ideal because: • Group work and managing interaction is difficult • You will have a difficult time seeing all of the learners in the room • It can be difficult to hear people's questions.	You may need to distribute microphones for people to participate so you and the classs can hear them.
U-shaped	*u-shaped style*	• Ideal for interaction between you and the learners. • You remain the center of their attention. • You can get close to each learner and maintain eye contact.	Not ideal with large groups.	Better layout to use for demonstrations or when using a screen.
Boardroom	*boardroom style*	• Best for small groups where discussion is the focus.	• Facilitator will always be behind someone in a full room. • No defined focal point in the room.	Most training rooms are set up like this by default.

TABLE 5.1 Room Design Options (cont.)

Name	Example	Pros	Cons	Comments
Classroom	*classroom style*	• Good for large groups. • Good for classwork or activities on paper.	Not much interaction between learners.	Make sure you know how big your classroom is before choosing your tables!
Banquet or rounds	*banquet style*	Is ideal for interaction between learners and small groups. Facilitator can move around the room and interact with learners.	• Can be used for large groups, but can accommodate more people using a different layout. • Some learners will have their backs to the facilitator or the content.	Table sizes can vary in this arrangement, double check the size you are getting.
Cabaret or half rounds	*cabaret style*	This provides a focal point for learners facing the facilitator or screen.	This limits the amount of people that can be in a room since only half of each table is used.	This is very similar to banquet layout, make sure to check table sizes again.
Herringbone	*herringbone style*	This arrangement can make smaller rooms easier to move around in.	Learners against the wall may not be able to move out of their seats with a full row.	This has all the benefits of a classroom setting.

design for safety

Design your room with safety in mind. I've had to teach in a lot of unusual places, but only a few pushed the limits of safety. Not only was I uncomfortable teaching in a space that was too small or allowed only limited movement, I didn't like the message it sent to the learners. If a room is too crowded, you can't move around freely or, worse yet, can't exit the room safely in case of an emergency. What message does that send learners? That safety is only part-time or when convenient? Tape down electrical cords, map out emergency exits, know the evacuation signals, and arrange tables to allow everyone to exit the room safely.

Making Training Stick

Remember when we talked about bridging the gap of knowledge or skill? Not only do you need to make sure participants learn what they need to, but you need to ensure it sticks with them. Almost all of your design and development should be with that goal in mind. Here is some more information to consider as you develop your content.

Different Learning Styles

The learning profession has had a lot of discussion on whether learning styles are a real thing. I am not going to get into that discussion; suffice it to say that we all learn our own way. Teaching how you learn will not work for most of your audience. You have to mix up all the different ways of delivering information so that everyone gets a little of what they need. Some learners want to talk about their experiences with the rest of the class, so plan for that. Some learners prefer just listening to what you and others in the class say and following along in their materials. Some learners might talk in their teams but never speak up in front of the whole class. Some learners can look at pictures, diagrams, or videos and learn everything they need. Some want to read ahead in the materials and "check in" during activities. Which of these learners does your class contain? Probably all of them.

Get Their Attention

One way to make sure learning sticks is to get the learners' attention and keep it the whole time. That's tough work, even for a seasoned ninja. It's possible, though, with good design and good content. Here are some more things to consider during your material development to make sure your training sticks.

- Play a game. Pick a game and use it to energize your class after lunch, after a tough technical session, or just for two minutes of fun.
- Explain objectives. As you facilitate your class, remember to keep tying back all your content to your learning objectives. It's very possible they forgot why they were engaged earlier, so remind them.
- Match objectives to their motivation. Ensure that what you are doing matches any motivation they have to learn. Review the section on ALP and be sure you are tapping into any motivation they have

shared during the class. In theory, everyone wants to go home safely, they just have to connect what you are teaching with that outcome.

- Put it in context. Use as many examples from their work as you can. Ask them for more so they are participating rather than you supplying all the examples.
- Link it to prior knowledge. Connect what you teach with any prior knowledge they may have. Use previous safety training, personal safety stories, and ask learners to share anything that is relevant.
- Get them curious. Ask open-ended questions to get them teaching each other, or share part of a story, prompting them to ask for the rest. Tap into any competitive spirit that may exist. Use treats or other items as praise to stir up participation.
- Vary the stimuli. Keep changing your style and delivery: lecturette, video, activity, reading, group discussion, flip-chart writings, question-and-answer session—just keep it varied, exciting, and relevant.

Chunk Your Information

Chunking is the concept of breaking down complex sets of information into memorable chunks. Some of you will remember when telephone numbers in the United States were seven digits. Most of us had the important numbers in our lives memorized easily. That is because most brains can easily recall seven plus/minus two pieces of information. With more people today and their cell phones, everyone now has a ten-digit telephone number. And guess what? For most of us, that is just one too many numbers to memorize. Now apply that concept to complex operations where safety is an important element. There is no way the training will stick—most brains just can't handle it. We need to chunk important technical information so that learners can remember it.

Your presentation should be clearly chunked. How many learning objectives do you have? Those are your first chunks. Often, while designing programs, we develop a chunk system for the whole course. In one supervisor safety course, we created the entire program in five chunks. We even designed the PowerPoint to have a graphic that reflected the five chunks visually. A great activity for your learners is to have them actually chunk the information for themselves. What if you had a sixteen-step procedure for the proper lockout/tagout of your mechanical equipment, can you expect your learners to remember those steps? On the other hand, what if you gave them the sixteen steps and asked them to create three sets of steps? If they create it in an activity, they are much more likely to remember the information, and the chunking of the information is a bonus that you can make into a job aid. Now that is being a ninja!

Make Lists, Tables, Diagrams

Use visual effects to help learning stick. Remember the activity I mentioned when discussing icebreakers that I use when I teach any Safety Training Ninja course? I know that if I flash the final image of the activity on a screen to anyone who has attended the course they will remember that image and the activity. Learners have come up to me afterwards and told me they will never look at that image the same way again! Tables and graphs help, as do bullets, graphics, and icons. This book uses bulleted lists and icons to help you remember as you read.

Use Mnemonic Devices

A mnemonic device is a tool for aiding the memory. It connects a new or unfamiliar idea to something that is known or easily memorable, such as a rhyme, acronym, pattern of numbers, and so on. You probably already know some, but you may not have known what they were called. HOMES helps us remember the US Great Lakes, and ROY G BIV is a way to remember the colors of the visible light spectrum (the rainbow). After researching, I discovered there are actually *a lot* of them. My favorite is from childhood. I was always confusing the two months in which the US holidays of Memorial Day and Labor Day fell. Now, I will never forget because someone told me May and Memorial both start with the letter "M." A great way to make a new mnemonic for your learners is to have them create it. The process of creating it will likely make it "stick" better than if you tell them what they should memorize. If it hasn't already popped into your head, the one most safety people shout out in class is PASS: Pull, Aim, Squeeze, and Sweep for fire extinguisher usage.

Tie It to Life Experience

This memory tool can be based on the learner sharing a memory that connects to what you're teaching, or you can trigger a memory with a story or image. This is when careful use of images is important. This ninja has never been a big fan of showing graphic examples of injuries to scare or repulse learners, but it works for some people. I think you need to think carefully about your audience and whether it's right for them. Just because some learners appreciate gory images doesn't mean they will not offend others. I prefer stories that pull at the heart strings over gory pictures.

Using Visuals

As we are completing our first draft, it's time to talk about adding a spark to your materials. But just a spark, not an inferno.

Images

Using images or having miniature models of what you are discussing can be a welcome addition for the more visual and tactile learners. As I mentioned, be cautious when using graphic images. What about images that invoke feelings or senses, perhaps connected to family, a job well done, or team work? I don't think every PowerPoint slide requires an image, but sometimes they help make your point.

Another important image consideration is to ensure that any work-related pictures you use actually relate to the work of the learners. Showing fall protection for steel workers won't resonate with residential home contractors. The wrong images can cause learners to shut down because they don't feel you really know your audience. Other images to avoid are manufacturer-provided sales images. If your learners don't have the latest equipment, the pictures won't work. Worse yet, they now know better equipment may be available, and you can have learning interference as they imagine having the newest tools. A little further in the chapter, be sure to review the section about ninjas having honor. Make sure you are using images legally.

Color

Some of you may be limited on how your content looks based on your corporate brand. If you can create anything you want, remember it's about the learners. Choosing colors that are pleasing to the eye and using fonts that everyone in the room can read may not be exciting, but this is best for the learners. Remember, it's all the hard work you do during design and development that will make the program exciting—not a curly font.

Icons

I like to develop unique icons for learning. You can reuse the icons throughout your follow-up evaluations and job aids. I often match the unique icons to the learning objectives to increase learning stickiness.

Videos

Make sure videos are clear when you use them. Have them expanded to the entire screen and adjust the video color and contrast in PowerPoint, if needed. Recently, I noticed a trend where trainers are putting their videos "inside" a graphic that looks like a movie screen or computer. It is clever but forces you to reduce the size of the video, and that is not in the learners' best interest.

Fonts

Stick with your corporate font or one of the classics, such as Arial. If you have downloaded unique fonts for your training development, and you use a different computer to deliver the training, the font may not transfer properly—or at all—and the PowerPoint could be a mess. This may mean reformatting your entire program at the last minute as learners start arriving to the training.

Transitions in a Program

As you are developing the training program, don't forget about your transitions. They are the part of the program when you move from one topic to the next, or for me, usually from one learning objective to the next. No one is ever fully engaged every second of even the best training program, and transitions help pull their attention back to prepare them for the next piece of information. You can be subtle in your transition, but this ninja prefers to make a transition a big deal. I usually announce, in an unmistakable manner, that we are moving to the next section, that we just covered XYZ and now are moving on to the next big area. I try to tie my transition to the bigger picture I've created, both visually and with information. This usually only takes a few moments, but it helps keep all the information the class is learning connected in their minds.

notes, notes, and more notes

The notes section in PowerPoint can be a great place to put all the extra content that you don't put on the slides. Any photos, credits, or references can also go there. If your training materials are audited as part of ISO or other standards, having your source materials in the notes will satisfy their auditing standards. Having all that extra material in the notes section is also a great backup if someone asks you a technical question based on what is on your slide.

Developing Your Training Materials

Developing Job Aids

You know that PowerPoint slide, the one with way too much technical information that they can't see in the back? The one where you say, "I know this is a lot of information, so I am going to make sure I talk about every step." Well, first, you know you should not have that much information on one slide, and second, if it's really that important, what else should you be doing to make sure the learning sticks? The answer is using a fabulous job aid! Grossly underutilized in the EHS world, probably because we are too busy with our regular jobs to develop "bonus" training materials, job aids can be wonderful! So, budding ninja, I want you to give them a try.

If you're asking "What are job aids?" I can tell you that you've probably seen them before. I had used them before I knew they had a name. Job aids include anything you can develop to support bridging the learning gap. They come in many forms, including:

- checklists
- worksheets
- flowcharts
- safe work procedures
- videos
- infographics
- pictures
- charts

FIGURE 5.2 Job aid for ADDIE

ANALYZE

The analysis phase is the foundation of a training program. This is where some pre-planning will serve you well. The basis for who must be trained, what must be trained, when training will occur, and where the training will take place is completed in this phase.

DESIGN

This is the road map for your training program. The outputs of the analysis phase drive the design and end in an outline of the training program for future development.

DEVELOP

This phase begins the creation of your program. We start to expand on the objectives and start filling in the outline. The "filling" is both the technical content and the methods of delivery.

IMPLEMENT

This is it, time to teach, train or facilitate. All your hard work and materials preparation are finally ready to deliver. If you mapped out the design based on your analysis, delivery will go a lot easier.

evaluate

EVALUATE

The last phase is to evaluate if you met your learning objectives and the materials you delivered will fix that knowledge or performance gap you found in the analysis phase.

- layouts
- conversions

Job aids support learning with extra or supplemental information to help you do the job safely. Not everything you need to share with your learners can be covered in the training, sometimes you have to supplement. Sometimes job aids are used by the learner with the assistance of a coach, supervisor, or you during an OJT safety moment. Other times, a job aid is used temporarily, until the skill or behavior is reflexive or confirmed by observation.

The key to developing a good job aid is to ensure that any directions for it and the desired outcome are easy to understand.

Remember the infographic in figure 5.2 (we saw it earlier in the book)? It's also a job aid for remembering ADDIE. Figure 5.3 is a job aid to help you determine if you should create a job aid. Clever, eh?

FIGURE 5.3 Should you build a job aid?

Writing Your Manual or Handouts

I am not a big fan of providing a lot of supporting materials unless they are in the form of job aids. If your learners need copies of a policy or legal requirement as part of any activities, of course these need to be provided on paper or electronically (and made accessible during class). If the new regulation is just for reference, save a tree and provide a link via a follow-up email or as a QR code on their final evaluation. I do think that many learners who take notes would prefer to take them on a version of the PowerPoint in front of them. The problem is, if you print one slide per page, it's too big to allow for notes, and if you use the default PowerPoint three slides with lines next to them, most people can't read them because the print is too small. I solved that problem by having a third party write a macro that converts all my PowerPoint slides into images and then into a table in Word. I use two slides per page, and they are much larger and easy to read. Next to the images of the slides is an open box for notes. See a comparison in figure 5.4.

The two-slide version allows learners to actually refer to the slide later. Now maybe you can't hire a third party with the budget you have (mine cost less than US$100), but what you put on your slides can affect how the final handout looks. If you've been following the rules for a good presentation, those three little slides on the page should be readable without a magnifying glass.

Whether you are writing content for your PowerPoint, a job aid, or a supplemental handout, keep these things in mind:

- Be smart in your design. We will talk about PowerPoint use more in chapter 10, but many of the same concepts apply for any printed materials. Use bulleted lists when possible, avoid long paragraphs of information that most learners won't read. Use drawings and illustrations to support written words.
- It better be right. Make sure what you write is not only technically accurate, but that you use proper grammar and have no typos. Sadly, some learners get so caught up in calling attention to errors that they miss all the great learning. Plus, it's no fun to get caught with an error. If I do get caught, I acknowledge the error. I ask the person to write down the page number where the error occurs and share it with me after class is over. This allows me to stay focused on what I am teaching, and as a bonus, I validate the ego of the person who

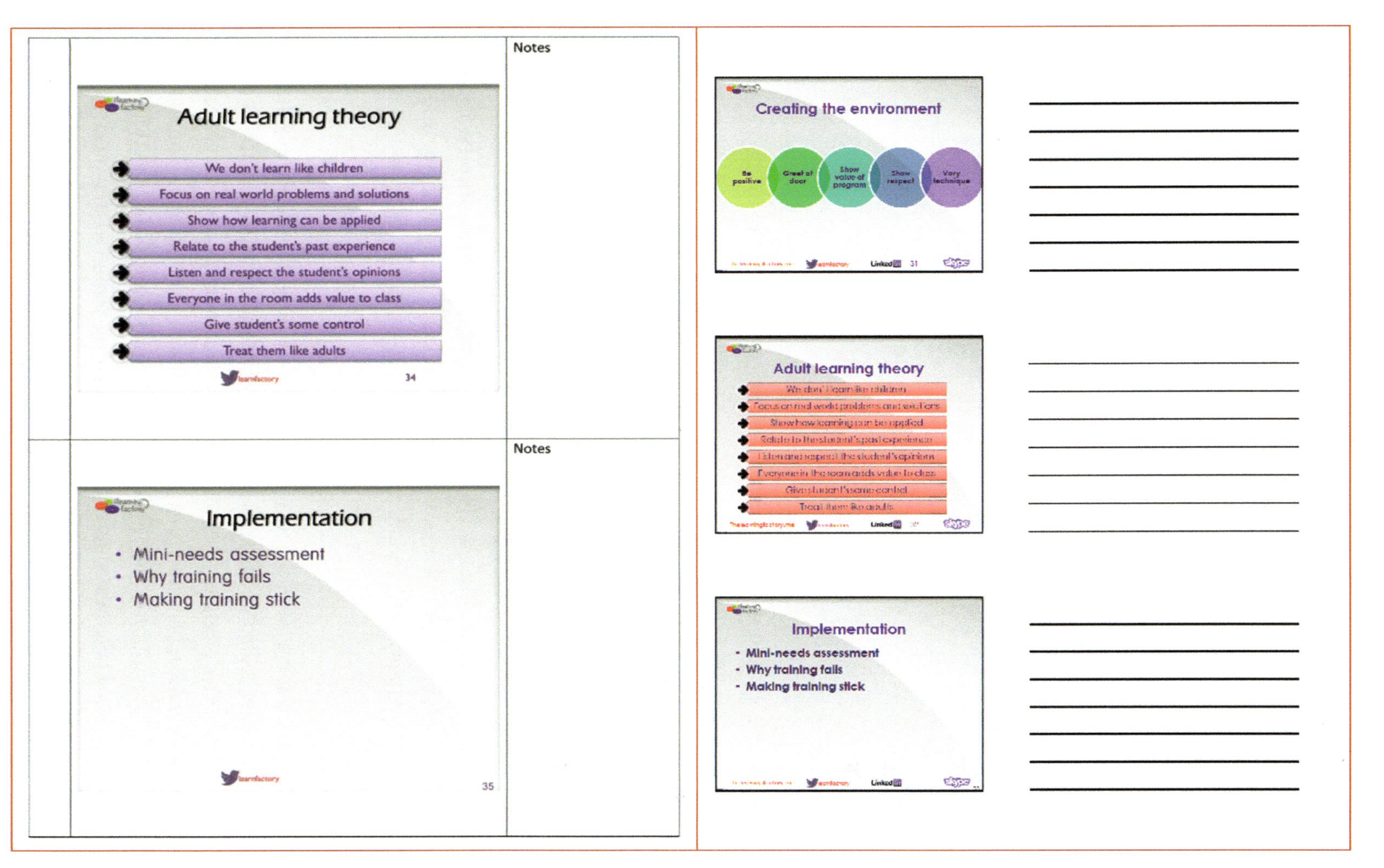

FIGURE 5.4 Comparing PowerPoint handouts: customized two-slide page on the left and default three-image page on the right.

points out the error—because that learner is out there, the one who needs to feel smarter than you. So let them feel smarter; it doesn't take anything away from you to let them feel smart.

- Make it relevant. We are back to the ALP again. The materials, PowerPoint, and activities must allow the learner to connect with the content. Relevance to their jobs, motivations, or safety all make the class more effective. I once taught a class where there was a revolt at the beginning of the class regarding the relevance of some of the materials. The class was almost hostile about it. I did not develop the materials, just showed up to teach the course, so I made a deal with the group. I would not talk about the topic if they forgave me when it kept showing up in the materials. Since it was not legally required information, or required to bridge the gap, it was an easy deal to make. After that, the class was great, and the learners were fully engaged.
- Make it appropriate for the audience. The words you use, both verbally and in print, must be audience appropriate: the right language, the right education level, and right for the industry. If you are training an industry you have experience in, it's easier to talk the jargon and share your experience as an expert. Use the information from your analysis to make sure you are writing and speaking the correct words in the correct way. Also, do some research on any cultural issues that may be relevant in your classes. If I teach in a foreign country, I do my best to stay within its cultural norms. Even if I am teaching in the United States, there are still cultural norms and globally represented workplaces. Take all that you know about your audience and design your content to make it work.

Closing

You have done a lot of work: written your learning objectives, developed your outline, filled in the space in your presentation with valuable learning information, designed your activities, created job aids, and developed your evaluation plan and forms. Now it's time to end strong. Don't just say, "And we are finished." Instead, build your ending using the ALP. Plan to recap the key takeaways, refer back to learning points shared, and remind everyone of the job aids or other sources of information they can go to

know your audience

Right out of university, I was teaching first aid/CPR to state employees. My boss was wise enough to tell me that our department had the highest illiteracy rate in the state and that many of our fine, hardworking employees could not read the test. He shared with me a tip that has lasted a lifetime. At the beginning of class, I asked, "Whose boss forgot to tell them they had training today and because of that you didn't bring your reading glasses?" Several hands were raised. I let them know it was no problem. I would just read them the test instead, they could answer me verbally, and I would fill in the information for them—no embarrassment for the learners and no surprises for me. What an amazing way to respect your learners by taking the time to learn who they are and what their experiences and capabilities include!

later. It is likely that the final evaluation or perhaps the personal commitment form mentioned before is the last thing to do. But be sure everyone knows they can stay after class and ask questions or connect with you later as needed.

Rehearsal

Unless you are already a content and delivery ninja, you need to plan on time to rehearse. Even the best designed and best developed programs need a dry run to ensure what you plan to do and what actually happens is the same. While the old way of training—reading endless slides to an uninspired crowd—may not have required too much preparation, you now have activities, open forums for discussion, and opportunities to let learners teach each other. That takes some finesse to schedule properly.

After almost thirty years of doing professional EHS training, I was reminded of this critical step in the development process. I was delivering a major program for my client. We would be rolling out the materials to more than twenty countries and hundreds of learners. The program was packed with activities and a lot of new concepts. Everything was set, and both the client and I thought we had it ready to launch. I knew my material very well, and all the timings had been tested, but my gut told me to force the dry-run rehearsal on the client—more money for him, but a lot was on the line. It was worth every penny; the dry run was an epic failure. It was possibly the worst teaching day of my career; everything was just "off." Within hours of completion, the content was torn apart, reworked, and a lot of material was removed or rewritten. We then rehearsed via conference call four more times before the launch. The first delivery was perfect: great participation, measurable learning, and positive feedback from the learners. Every class thereafter was better because we used feedback from the evaluations and ADDIE cycle to improve content and activity flow. If we had not done the dry run and further rehearsals, it would have cost my client a lot more money and, worse yet, his reputation for both hiring me and not delivering on the promise he made to his stakeholders.

Ninjas Have Honor—Don't Steal Content

Regarding the issue of using copyrighted materials and images, do not steal content. Just don't do it. It's theft, plain and simple. If you work for a company that has any assets, you are setting them up for legal action. I myself have had materials copyrighted over the years and worked hard to develop them. I can respect why others would not want their hard work stolen for someone else's benefit. Here are a few thoughts on the copyright issue:

- Just because it is on the internet doesn't mean you can legally use it. The internet is filled with copyrighted materials—think online

design your flip charts as you design your content

As you design your activities, think about what props you will be using and how they will be used. One great trick is to design your flip charts in advance. Sketch out what they look like on paper and have that example in with all the training materials you will be taking to class. Knowing what the group is going to do when they stand up during a brainstorming session or group activity is just another element you can plan for to make your EHS training great. This will also help to keep your stress level down as you teach. See the next chapter for more flip-chart tips.

newspapers, videos, blogs, and private companies' content that is used to generate business. Those are not yours for the taking.

- Just because you can tie the materials to a government website does not mean you can use them legally. In the United States, many of the government's materials are categorized in the public domain, but not all. Unfortunately, they don't always indicate which ones you can use. Here is a simple example: If you go to a US government website and see a great picture of Mount Rushmore, you might think the photo is public domain since your tax dollars funded that site. Probably not. It's likely that the photo was taken by a professional photographer, who sold the use of the image to the US government. The photographer owns the copyright, not the government. In regard to other government entities around the world, you are going to have to research whether you can legally use their content.
- Just because it is on the internet does not make it reliable, good, or even correct. Be careful what you copy and use. There are endless examples of EHS training topics and materials to be reviewed, but just because someone has the skillset to upload content to a website doesn't make it accurate. Use the internet as an information and inspiration source, but don't steal.
- Just because you don't think you will get caught doesn't make it OK. Hey, isn't that an unsafe behavior excuse we really dislike hearing, "I didn't think I would get caught"?

So, when is it OK to use others' materials?

- When you have permission from the owner. Send an email to the owner of the content and ask for permission to use it. Explain exactly who you are and how it will be used. Most of the time when I do that, I get an email back from the owner granting permission, provided I give the owner credit. I am happy to give credit; that's easy. I copy the contents of the email and place it in my slide notes so the record of the agreement is with the content.
- When you've paid to use the material. Publishers will often require a payment to license materials that have been published in their books, especially images. Weigh the uniqueness of the material against what it will cost to use, as some of these fees can be in the

hundreds of dollars per image. You can also license images from stock photo websites to use in your presentations. Read the terms of the agreements carefully to determine if you have unlimited use or only a one-time use.

- When the material is in the public domain. As mentioned above, this can include government documents (but not always, so be careful). There are also many websites that offer images that are in the public domain and free to use. Sometimes crediting the photographer/author/provider is necessary but often not (though it would be a nice thing to do and appreciated by the provider).
- When your use of the material is considered fair use. This can be tricky and hard to define. Excerpting brief passages of text may be considered fair use, but images rarely qualify, except in specific instances. Be on the safe side and ask for permission for any copyrighted work. When copyright is involved, asking for forgiveness is never easier than asking for permission.

Generating Interest in Your Training Program

I coach budding ninjas to make sure they market and advertise their programs. Some EHS professionals will cringe at this idea, so get some help if you need it. The work will be worth it. Consider the following to introduce not only a new training class but to capitalize on the new way you are using ADDIE and ALP.

- Create an email to the learners explaining the learning objectives—WIIFM—and how the class will be different from past EHS training. Include any prerequisites or special PPE they may need to bring for any shop or field exercises. Be sure to include all the logistics: date, start and end times, who should attend, etc.
- Create posters/flyers that generate excitement about the course and post these in the break room, job trailer, or at elevator bays so everyone can see what is planned. There are plenty of websites that offer free tools to develop a quick and easy advertisement based on precreated templates.
- Walk the floor or head onsite and talk about the training at a morning huddle or toolbox talk. Bring along some candy or trinkets the

learners might like to create a positive link between you and the program.

- Advertise the facilitator, not just the program. If you or one of your facilitators is well regarded by your learners, leverage that with a picture of the facilitator being safe or teaching a class.
- Use your company's intranet to tweet, yam, or blog about the upcoming program.
- Make an exciting short video of the facilitator, subject matter expert (SME), or a learner who helped in development talking about the program and WIIFM. You could host it on your intranet and post QR codes in strategic places. Add a flashy note that says, "Watch me."
- If your class will be taught several times, make a short video of learners discussing their positive experience after the program. Post those anywhere people will see them.
- Can you think of other ways to generate interest in the program? Remember, your goal is to deliver great training to bridge the gap of knowledge or skills. It will be a lot easier to do if learners are excited about coming to the class.

Z490.1 Supporting Material

As we close out the chapter, I want to reference several locations in Z490.1 (ANSI/ASSP Z490.1-2016) that support much of what you have just read. Section 5.2.1 suggests delivery methods and materials that include:

- proper planning and preparation to deliver
- methods to manage the learning environment (more on this in chapter 6)
- effective use of PowerPoint, videos, and other mediums and technologies
- using ALPs that are right for the target audience
- communication methods for feedback

One of my favorite parts of the Z490.1 standard is about managing the physical learning environment. Section 5.2.2 speaks to the safety of learners during the course. This book and Z490.1 are a couple of the few places where you will see that idea in a training creation model. Z490.1 states that the location of the training needs to:

- be suitable for the learners needs and be prepared before training begins
- be safe and free of obvious hazardous conditions
- be quiet enough so learners can actually learn the materials
- have access to water and restroom facilities (did we really need to say that?)
- have proper indoor air quality
- have proper lighting for reading, writing, and learning
- consider ergonomics for training activities
- have a planned evacuation route and adequate emergency exits
- have a way to summon emergency medical services

ANNEX B.10 reminds us to be sure our training aids are right for the content and delivery. Review the following for more support on choosing the appropriate learning aids.

- Use a variety of training aids, choosing them to support your learning objectives and the activities you have designed.
- Choose the types of aids you use based on the audience and the knowledge and education levels of learners.
- Balance the different types of aids and tools you use, making sure they support your learning and business objectives.

Annex B.13, Develop Training Materials, also supplements many of the key ninja points, including:

- Create a trainer's guide that documents all the course materials so the course is reproducible by someone else. Include learning objectives, course outline, and plan of instruction for the entire course.
- Review your plan for format, stakeholder brand, sequential organization, and accurate, up-to-date materials that support ALP.
- Include visually appealing materials that are relevant and easy to follow.

- Properly include reference materials, usage approval for copyrighted materials and images, and the location of reference materials.

ANSI also recommends a written guide for trainers delivering the materials. This is very important if you are designing materials for others to deliver.

References

ANSI/ASSP Z490.1-2016. 2016. *Criteria for Accepted Practices in Safety, Health and Environmental Training*. Park Ridge, IL: American Society of Safety Professionals.

Biech, Elaine. 2015. T*raining and Development for Dummies*. Hoboken, NJ: Wiley.

Chapter 6
Implement Your New Training Program

You have done a lot of work, and now it is time to deliver your new training program. Here are some things to remember before that big day.

Before You Arrive

Consider these basics to make sure you are not stressed out before the class.

Confirm Your Location

Speak to your host or the sponsor of the location a few days in advance. I have been kicked out of my training location plenty of times by a "more important" person or group that needed the space. With a few days of warning, you will have time to recover from a complete change in location and be able to adapt as needed rather than doing it on the fly.

Plan How You'll Get to Your Location

Map out the best way to get to the location. Are you driving or walking? How will you get all your training materials to the classroom? Do you need a hand truck, dolly, or maybe a helper? If driving, find out about road construction that could slow you down or cause a detour. Also check your phone's map app in the morning to see if there are any traffic accidents that might cause rerouting or require you to leave earlier.

Make sure there is parking available for you and your learners. On one occasion, I arrived in plenty of time to set up; however, the parking garage was full and there was no one there who could help me. Class ended up starting late after several apologies. Once I spoke at a conference that was "right across the street" from the hotel, according to the hotel staff where I was staying. The walk ended up being about fifteen minutes—obviously, it was not across the street. Luckily, I planned for the extra time. But I did not plan for it to be dark when I began my walk, which was further hindered by snow that fell the previous night. Halfway there, I fell on black ice covered with snow, landing face down. I threw my coffee all over the ground and hurt my hands and knees. The walk ended up being closer to twenty-five minutes because I had to slow my pace and was afraid of falling again. By that time, learners were starting to arrive. I flagged down a truck with the company name on it and got a ride the rest of the way. Not the best way to start the day!

Tech at Your Location

Connect with the IT person onsite. Will they be there as early as you? If not, you might not have anyone to help with technical issues. Find out the internet password in advance. This is critical if you are accessing online information during class. If by chance you can't find your presentation, you may be able to access it online for a quick download. I always try to have a special thank-you treat for the IT staff who show up early. You never know if you will need them again, and you're paying it forward for the next person.

Early Setup of Location

Set up the space the night before if possible. I get a much better night's sleep if I know I have nothing to worry about but the final A/V hookups and whether the coffee arrives on time.

Breaks at Your Location

Find out the timetable for any breaks or meals. You want to know when the coffee will arrive in the morning and if there's a refresher? What time are any meals or snack breaks planned for? This is critical information for you and your class. Providing a food and coffee schedule is one way to show your respect for your audience.

Facilities at Your Location

Know where the break area, lunchroom, bathrooms, smoking areas, and emergency exits are located.

Use the timing breakdown in table 6.1 to remind you about when preparations closer to training delivery should be done.

TABLE 6.1 Checklist for Class Preparation

Timing	Task	Date Completed
At least two weeks before the training	Determine any last-minute learner needs, priorities, and prior knowledge that may need to be considered in the training delivery.	
	Verify training outcomes and a final agenda.	
	Determine the final number of teams/learners.	
	Confirm the training location and room setup.	
	Ensure training materials will be printed and ready.	
	Review material and practice presenting the material.	
Two days before the training	Finalize the training schedule with stakeholders, leaders, and learners.	
	Gather training supplies, white boards, flip charts, markers, etc.	
	Confirm start and end times.	
	Confirm all food and beverage catering.	
	Send a reminder email about the upcoming training.	
The training day	Arrive at least 60–90 minutes early to set up.	
	Ensure presentation and technology is working properly.	
	Test videos and the microphone to ensure the audio is functioning.	
	Confirm break and lunch times.	
	Place supplies (e.g., markers, sticky notes) in the center of each table.	
	Check that all participant materials and sign-in sheets are available.	
	Ensure the room layout maximizes team discussion and the viewing of slide content.	
	Introduce yourself to learners as they arrive.	
	Gather completed training evaluations, commitment forms, or other documentation.	
Soon after the training	Analyse participant training evaluation data and compile the results.	
	Share evaluation results and discuss next steps with the stakeholder.	
	Follow up on any unanswered audience questions.	
	Follow up on any procedural omissions/issues identified in the training.	

Whether you are packing for a trip by airplane or a walk down the hall, use the checklist in table 6.2 to remind yourself of all the tools you may need.

TABLE 6.2 Checklist of Tools and Materials

Item	Quantity	Who Is Bringing	Notes
Electronics			
Laptop			
LCD projector or similar device			
Screen			
Extension cords			
Adaptors for electrical devices or LCD projector			
Power strips			
Audio speaker with extension			
Electricity adapter			
Other training mediums			
Flip charts			
Easels and stands			
Markers			
Computer files			
PowerPoint files			
Audio/video files			
Links to backup files			
Learner's materials			
Handbooks			
Activity directions			
Worksheets			
Table tents			
Pens			
Notepads			
Sign-in sheet			
Miscellaneous			
Treat bag			
Treats, candy, sweeties			
Lip balm			
Medications for you			
Hand lotion			
Business cards			
Other			

Room Setup

Although you should have planned your training space during design and development, you will always need to check the space out the day before or get there early enough on training day to fix anything that's out of place. Training rooms often become the storage space for lots of unwanted items, which means you may need a few hours to clean up and get the right arrangements for your learners' tables and chairs, your screen, LCD projector, flip charts, and any other tables or extra space you need for props, equipment, or team-working areas. I have walked into many spaces that were set up for me in advance, only to find I needed to rearrange things to match the training program design. This ninja always arrives at least ninety minutes early to get my room set up and to relax and get into my training frame of mind. If you don't use all that time, you can chill, knowing that all crises have been averted.

I remove any unwanted tables and any extra chairs so that learners don't sit there accidentally. If the room is large enough, I try to create an active space where I will teach. As the learners start to arrive, I let them know they will be filling the available tables completely. A room that is too large can impact a class culture by promoting a sense of disconnection from you and the other learners, and if your learners are really spread out in a high-ceilinged room, it will be hard for you to develop a connection with them and vice versa.

Create Your Space

I bring a lot of stuff with me when I train, and I need my own table for all of it. I need to create my space to look like all the other spaces I have created in the past. It gives me comfort and confidence that I have thought of everything. I request a table at the front of the room away from the screen. I put everything I will need on the table and start to set up my A/V and the other equipment. During my classroom design, I make sure I have space on the table for copies of all my materials, my treats, candy or sweetie bag, a bottle of water, lip balm, and throat lozenges. If needed, I have some headache relief right there as well. If I am using a lot of props or demonstrating equipment, I make sure I have a place to conveniently access them all. My space ensures that I can relax after I have set it up. This frees up my time for greeting my learners as they arrive.

Electrical, A/V, and Your Space

During your development phase, you should have determined what equipment you need to teach your class properly and how many power outlets you'll need. You should include the need for power strips, extension cords, and any A/V requirements. If you are in a room with complex lighting systems, work with the building control team to learn about them, or ask for a cell phone number in case you need their help. Also, take advantage of anyone in the class who may be familiar with the space. One of my classes had the most confusing light switches I have ever seen. I could not master them, but one of the learners held his weekly team huddles in the same room. I kindly asked if he would act as the lighting expert for the rest of the class. It worked out great.

If you plan on showing a video, you need to run it in the room ahead of time to check both the visual and audio quality. If speakers were provided, do they work and can the audio be heard in the teaching space? Can you adjust the audio easily or will you need help from IT? Be sure you check all the buttons, speakers, controls, and cords long before any learner arrives. Make sure you have all the electrical cords taped down to avoid tripping. EHS learners love to point out safety hazards in your classroom setup!

As technology advances, I struggle to keep up with what is offered around the globe. There are some amazing teaching spaces that even the building owners don't know how to operate. Make sure your computer runs any of its diagnostics before class starts. My virus protection comes on when my computer starts. I want it that way since I travel so much, but that means that it needs to finish its check before my instruction begins. Just another reason I insist on being there early. I also make sure I am logged out of everything but what I need to show on the screen. My email is closed; I am logged out of the conference call software; I don't have instant message programs open. If I don't do this, something always seems to open at the most inopportune moment.

Ensure there is a convenient space for your laptop and LCD projector that allows your delivery to be smooth. If you are using a microphone, make sure it works and is located where you can teach comfortably. I cannot teach from a podium; I move around a lot as part of my engagement strategy and general energy level. I don't like to wear lavalier microphones (lavs) because I can generally project to at least two hundred people (set up theater style), depending on room dimensions. What I am wearing can cause interruptions with a lav because the clothes and the lav rub against each other, which is a

browser history

Keep your computer history private by clearing the history the night before you teach. If you plan on accessing the internet during the class, open each site in a separate tab and have them ready before you teach. No one needs to know what website you have been on recently, and it could cause some class distraction. If you plan on going to Youtube, clear that history as well. It is best to open all those locations in advance and go to the tabs as needed. Don't attempt to find websites during class; it can waste time and be distracting if you can't locate what you planned to show.

nightmare for anyone listening. The lav's battery pack must clip to something. If I am not wearing something that has a place for that clip, that can be a problem. If I am being recorded or the room setup or size requires my using a lav, I do sound checks as soon as I arrive and again right before I start. I take the lav off after the first sound check and keep it off until right before I begin. If you forget you have it on, you may be sharing private conversations with the whole room, or worse, you may be wearing it when you take your final bathroom break before you start the class. Yes, it happens to the best of ninjas!

A Ninja Can't Control Everything

Once I was training in a new building that was having final equipment installation in the downstairs factory area. The upstairs training room was a lovely, well-lit space, set up exactly as my stakeholder had requested it. The ladies' restroom had no running water, but I could easily go downstairs to another restroom. About two hours into the class, the stakeholder excused himself from the room, and five minutes later returned and informed us that we had less than ten minutes to vacate the facility. The building had failed the fire safety inspection the day before and had no working alarms or fire suppression system as required by local building code. The construction manager didn't think to share that information with anyone, and we didn't know to ask, assuming there was a certificate of occupancy. We had to move to a very small space in a matter of minutes, carrying all of the materials and equipment to the main building. The learners were very understanding about the crowding in the new space, though the stakeholder did not take it as well. Being removed by the fire marshal for unsafe occupancy during EHS training can make it hard to maintain the desired safety culture.

Preparing for Distractions or Learning Interference

Even the best-planned EHS training course can get bogged down by distractions. Getting angry or feeling offending by learning interference only frustrates you and your learners. The best you can do is accept the challenges of today's workplace and teach who is in the room the best way you can. Drowsy learners, hostile learners, bad weather—it doesn't matter. What matters is how you react to the challenge. I promise the better your preparation, the easier the distractions will be to handle. Here is a list of some and what to do about them.

Equipment Operating Nearby

There is nothing more exciting than watching heavy equipment operation, even if we see it every day. If it is happening right outside your training room or trailer, just acknowledge it—let the learners take a few moments to watch and get the interest out of their systems. You can't fight it, so embrace it.

Bad Weather

A serious rainstorm or snowstorm can distract your learners, especially if it looks like they may need to leave early or are worried about family members' safety. Use the next break to speak with your stakeholder, brief them on the latest weather forecast, and see if adjustments to the schedule are needed. If it's just another stormy day, you may just have to talk louder.

Room Size

We often don't control where we teach, but getting there early at least helps you adjust to too large or too small of a room. Rearrange the room as much as possible to make it easier for the learners to move. Three adults at a two-person table will make your room temperature rise, and it could make for some unpleasant proximity with a stranger. If the room size is a problem, discuss it at the start of the class. This lets the learners vent their feelings in the beginning, and then everyone agrees to move on.

Room Temperature

If you are lucky, you can adjust your own room temperature. But that doesn't mean everyone will be happy. If you have experience with your training room and know it will be hot or cold, advise your learners to dress light or to bring a sweater. Again, acknowledge the issue and ask everyone to move on—especially if there is nothing you can do about it.

Too Close to Their Work

If learners can see or hear others performing their work activities from the training room, they will likely be distracted by what is happening without them. Getting them back from breaks and lunch can also be challenging. Remind them that staying on schedule means everyone finishes class on time or maybe even early.

Chairs

An uncomfortable chair can be the worst distraction, especially if your learners are used to being on their feet most of the time. Encourage them to get up and stand if they need to, walk around the room to get their blood flowing, or do whatever it takes to keep them engaged.

Shift Work

If you have learners coming off the third shift with no sleep, it will be a challenge to keep them engaged. I like to ask if anyone is in that situation or if there is anything else that could cause drowsiness. I am often told about new parents who are not getting much sleep. I do my best to be both understanding and engaging.

Electronic Devices

FOMO stands for the "fear of missing out." Usually the term is used for people who worry they are missing important information on social media or through interactions with others. It exists in the classroom as well, with people worrying about what they're missing while in class. Cell phones, tablets, and laptops can be an enormous distraction to the user and the other learners.

I don't allow the use of laptops or tablets unless everyone is using them for the class or note taking. One or two learners using them while everyone else tries to be engaged is disrespectful to the other learners. It's even worse if a supervisor or manager is doing it. I usually ask the stakeholder sponsoring the course to speak to them about closing down the computer during instruction times. If the stakeholder is not around, I whisper discretely that they need to respect the other learners and turn off their devices until break time. Usually they smile and say, "Of course. I'm sorry." If they explain about an important deadline that must be met, I ask them to excuse themselves from the room until they are done. They usually come back and are often an eager participant upon their return.

Production Timing

I always encourage my clients to schedule training right after the end of the month or after a big production deadline. If your learners think they have to choose between the learning objectives and keeping their job, they are going to lean toward keeping their job every time. You can make them sit in the room, but you can't make them learn. I often have learners who let me know they need to leave class but will return as quickly as possible. I appreciate that information and thank them for it, though it's better to have scheduled the training so learners don't need to make those choices.

Current Company Culture

If you're unlucky, a day will come when something happens at the plant or on the jobsite that cannot be ignored. The worst-case scenario would be if someone is seriously injured or killed onsite while you are training. This news spreads quickly, so seek out your stakeholder immediately for any insights and determine if program adjustments are needed.

ninjas should be caring

When I'm teaching new ninjas, I like to share this story from a colleague. He tells anyone in the class who may be feeling drowsy to walk around in the class, and if that doesn't help, to go ahead and lay their head down on the table. Is that training blasphemy? It is unlikely that a learner is retaining much content if their head is bobbing up and down. Why does my colleague tell his learners that? Several years ago, a learner, coming off third shift with no sleep, fell asleep in class. His head fell forward, and he smashed straight into his metal coffee mug. The learner was injured and bleeding, and, ironically, it was the first on-the-job injury the company had experienced in a long time. Safety training should not be dangerous. It's wise to care for your learners.

Another distraction for learners is when they receive news or emails about the company's financial health. In the middle of one class I was teaching, I noticed a distinct change in the room as everyone looked at their phones. The company had just announced on the intranet that no bonuses would be paid that year. I asked how everyone felt about the news. They indicated they expected it, so I said, "Ok, then let's keep learning so we have a great, safe next year."

Be a Ninja from the Moment You Arrive

You have done all your preparation and now it's time to conduct your training like a ninja. Before you dig into the meat of the training program, here are some items to help with the beginning of the class. You want to get off to a solid, positive start since it sets the tone for the rest of the training.

Presenting Yourself

If you know your audience, then you know how you should look. Training executives may require more formal clothing, while field or jobsite training may require you dress for that environment. If you are teaching in a classroom but have exercises that require PPE, be sure to bring it with you. I always carry hearing protection and bifocal eye protection so I know that I have the kind I need to set a good example. If your audience is unknown, I always dress at least one level up from what I think the learners will be wearing. You need to present yourself as an expert and a professional; find out as much as you can so you can dress the part.

Greet Them at the Door

Even if you know your learners, greet them as they enter. Set the tone for the class, letting them know it will be informative and that everyone should participate. If you create a positive class culture from the moment they arrive, your good attitude will spread to the learners. I use that time to get a feel for the mood of the class and perhaps gain some insights about the current company culture. This meet and greet time also helps you overcome any nerves you may have about training in a new way. A colleague who attended a one-hour ninja training program emailed me recently on his success with greeting learners as they arrived. He was guest lecturing at a university, and he took the time to connect with the learners as they entered the class. The reviews he received thanked him for that behavior because the learners felt respected and included. What a simple way to make your training succeed!

Show Them It's a New Day for Safety Training

If this is your first time employing ADDIE-inspired training with activities and interaction, make sure you let the participants know they are in for a pretty great day. Explain that it won't be all lecture, that their opinions matter, and that you want to do all you can to make the time valuable and relevant. When I tell learners that the class will be unlike some of the other EHS training they have been to in the past, they are skeptical. But when someone comes up during break and lets you know they have never learned so much or had so much fun in EHS training before, all the preparation you have done will be worth it.

Explain Materials and Handouts

Right after I finish the icebreaker and have everyone energized, I quickly explain the materials they have in front of them (the agenda, training materials, handouts, and the PowerPoint slides), and what the class schedule looks like, and how PowerPoint will be one of many instructional tools I will use. This helps manage learners' expectations, sets the tone for some ALP, and allows for questions—especially the all-important ones: when are the breaks and what is for lunch?

Table Tents

I am a big fan of table tents (or name plates for our European readers). That's because I am terrible with names and want to be sure the learners know I respect them and want to address them by name every time I can. I bring lots of colorful markers and challenge them to be creative with what they write or draw on their table tents. An interesting phenomenon of class culture occurs when one person starts to doodle. It starts a chain reaction, and almost everyone else starts to add to their IDs. I always see this as a sign of the learners' comfort and confidence. On top of that, I get to know them even better through their doodles.

Flip Charts

A great trick I cannot claim as my own is to prepare your flip charts before the learners arrive. Any lists, charts, or questions the class may need to answer can be prewritten on a few sheets behind the first blank sheet of paper. Then, when you need to show how to complete an activity, you already have a sample done or the framework completed. This ensures that everyone is doing the activity the same way, which keeps your class running smoothly. Be sure your flip-chart markers are for paper and not for a white board. Paper dries out white-board markers quickly, which leaves you nothing to write with. If you use sticky flip-chart paper for attaching to a wall, it should be several sheets deep so it doesn't bleed through. Hotels may charge a fee if their walls are damaged, or your facilities management team might not let you use the good training rooms again. As I mentioned before, I always bring markers with me, just in case the stakeholder doesn't provide the right kind.

Activities

Put in the effort to achieve excellence. You have worked hard to design and develop your activities. Make sure your directions are clear before you release the teams to complete activities. For the more complicated activities, I might show a simple example of what I want from them when they are done. Prepare that flip chart in advance if that makes sense.

Take Charge of the Room

Someone decided you deserve to be the person up in front of the room facilitating this class. Stand tall and exude confidence. If you have followed ADDIE, rehearsed, and had plenty of time before the learners arrived, you should be ready to stand tall and take charge of the wonderful product you created.

Trainer Versus Facilitator

I have avoided the discussion of what the difference is between the two and have used the two terms interchangeably. They are close enough in definition for our work in EHS, but I do like to think that a facilitator allows for more flow between the learners and lectures less from the front of the room. A great facilitator knows when to be quiet and let the class teach itself. When a learner has an "aha!" moment and shares it with the others, a ninja stays quiet and lets the learner take all the credit.

Sign-In Sheets

Finally, don't forget the all-important sign-in sheet. If you are training for the company you work for, you know its policies and procedures regarding attendance. Otherwise, make sure you ask your stakeholder what and how you can ensure that credit for attendance is recorded.

Conduct a Mini-Assessment for Your Learners as Class Begins

No matter how extensive and complete your assessment and analysis are, it is always a good idea to conduct a mini needs assessment at the beginning of each session. "What are your hopes and fears for this session?" "How much experience do you have with . . . ?" This does two things: first, it helps determine whether the design is on the mark; and second, it gives learners an opportunity to state their expectations and to start participating early. You can also use what you learned from the icebreaker. If time is limited and you are unable to conduct a mini-assessment, you may still want to gather some data. To do this, speak with your stakeholder, talk to learners as they arrive for the class, e-mail a simple questionnaire in advance if you can, or contact other trainers who have worked with the learners before.

Managing Them

So now it's time to get to the body of your program. Almost all my training looks the same in the beginning: the title of the program, emergency procedures for building evacuation, the icebreaker, any rules of engagement for the learners, and then the learning objectives. After that, you just keep going. Here are a few items to consider for the opening as well as the rest of the program.

Rules of Engagement

A lot of my stakeholders like to have a slide for the emergency action plan (a great idea) and one for the rules of engagement. Simply put, these rules describe how the class culture should grow throughout the day. Examples on what to include on this slide are: silencing electronic devices, encouraging participation, respecting other opinions, speaking freely on how to improve safety, and returning from breaks on time. There may be other rules you want to include. Whatever they are, be sure they are right for your audience.

A Participation Pact

After I have done the icebreaker and gone over the rules of engagement, I try to get an agreement on participation in class. Without hesitation, I ask the question: Are you willing to open your mind to a new way of learning in order to have a great training class? Almost everyone says yes, sometimes eagerly, sometimes because everyone else is saying yes. This also builds the positive class culture for your training.

First Candy or "Sweetie"

I use rewards and treats as hooks throughout my training classes. I first saw this done about five years into my career. Two professional trainers had candy in the cutest little bag. It was genius! I started carrying candies and treats after seeing that, and I entice learner participation with them to this day. I make a big deal about the first treat, speak louder, build up excitement, and make sure everyone in the class knows that they are just a comment or input away from their own treats.

Recently, I didn't have time to stock up on my usual candy and had to buy some at a convenience store. It's hard to find a nice variety of individually wrapped candies at a convenience store. But I noticed some interesting lollipops; they were mango-flavored with a chili powder coating. I grabbed two bags. They were a big hit! Handing out treats is not about how amazing the candy is, it's about creating competition among the learners, about rewarding behavior that promotes learning and participation.

I learned quickly, while teaching globally, that even the simplest mistakes have an effect. Many Europeans call candy "sweeties." My offering of candy as a reward or treat was not going very well until I started calling it by the correct name. I kind of like using the word "sweetie." It just seems more whimsical. I have also given out rubber duckies and ninja toys as treats. Ninety-five percent of the audience loved them and even made the duckies squeak in support of others' good comments or suggestions. The other five percent thought they were childish and demeaning—a great example of how it is almost impossible to please everyone. Try to determine what is the right treat or candy for your audience and see if the concept works for you.

Humor

If people tell you that you are funny, and you hear that a lot, you probably are! But, if you are not funny, then using humor during training, especially EHS training, can ruin all your hard work. A dry sense of humor doesn't tend to work in safety training either. If you are not sure you can use humor effectively, ask a trusted colleague for an honest opinion. Try not to get upset if what you hear is not what you expect. Better to leverage your strengths.

Your Pace

EHS training is hard work and you must do everything you can to keep your learners engaged and focused on the materials. That means you may need to up your speaking pace to keep them tuned in. This might be a real challenge for you. It was very difficult for me when my speaking pace slowed down while teaching in foreign countries. I was not perfect at the pace change, but I never stopped trying.

Using Job Aids

Don't forget to use your job aids, and make sure the learners refer to them as you explain them. If I think I might forget to mention the materials, I put a secret symbol on the PowerPoint slide that only I know about. When I see the symbol, it triggers my memory to direct everyone's attention to the job aid.

Using Handouts or a Manual

If you have them, refer to them often. I like to ask the class what page I am on. It is a simple test of their engagement. One of the best sounds in the training class is when you hear the whole class turn the page of the materials at the same time. Now that is engagement!

Adjusting Your Delivery

Despite all you hard work and planning, something will impact your schedule and cause delays. Be ready to pare down content or shorten your activities to ensure you end on time. Alternatively, what if you are not getting the participation you planned on and are way ahead of schedule? I usually have to fight the urge to drag it out. My client expects a full day of training, so that is what I should provide, right? Not always. I speak to the stakeholder and explain what is happening. They may suggest ways to engage or support ending early. The key here is not to drag the day out with random fillers just for the sake of meeting a schedule.

Breaks and Lunch

Confirm that breaks and lunch are on the schedule and what you planned for. If the schedule has changed, be ready to adapt. If lunch will be served during the training, be ready to break for lunch the moment it arrives, even if you are not at your planned stopping point. Once learners see and smell food, that is their only focus. I have watched heads turn and eyes follow the food cart as it is brought in. Accept it and take the break.

Encourage Participation with Questions

This should become an active part of every delivery from a toolbox talk to a forty-hour refresher course on hazardous materials. Getting the learners to start asking questions, answer open-ended questions, and engage each other in learning is a critical element to ALP and your success. If you are having trouble getting learners to engage in your new style of instruction, use a break to your advantage. Seek out an informal or formal leader in the class and ask for help in getting the class to become more engaged. Often, that learner may volunteer to take the lead or may have good advice on what you can do to engage the class better.

The Difficult Learner

This learner goes by a lot of different names, and you have likely met a few in your experience. Here are some ways a ninja can manage this learner during class.

- Grumpy/angry. If this learner is actually disruptive, speak to him during the break. Ask if there is anything you can do to make the class better for him. Sometimes the learner doesn't even realize how toxic his behavior is. If he won't improve his attitude and it's affecting the other leaners, ask your stakeholder for support. Don't engage his attitude in front of the class. No matter how bad his behavior is, you will look worse because you are the one in charge.
- Class clown. Let a few jokes pass until it becomes disruptive. At the break, share with the learner that you have a tight schedule, that you want to ensure everyone gets to leave on time, and that although you appreciate making the class enjoyable, the jokes are slowing down the delivery. Ask for a commitment to control the jokes. Most people will stick to their commitments when you ask them directly.
- Know it all. We have all had one of these. Most of the time this learner is looking for validation for what he knows. So give it to

handling overzealous sharers

Some people just want to help teach the class. This is great if you know it in advance and plan for it, but what if someone won't stop sharing? If I can't change the behavior with an intervention during a break, I will usually change the way I ask open-ended questions. I say something like, "I want to hear from someone who has not shared yet. So, if you have been super helpful so far, I appreciate it, but I am only taking answers from those who haven't shared yet." Oddly, the quiet ones will start to speak up. In some cases, they were letting others do all the answering because it wasn't worth trying to be faster to speak up.

him—shower him with praise for a contribution. Often this is all that's needed. If the learner becomes disruptive, follow the same step as before: at a break, ask him to help finish the class on time by limiting how much he shares and give others a chance to speak. Still doesn't work? Seek out your stakeholder for intervention.

- Silent type. Don't assume this person is not engaged. Some people just don't talk in class. Watch this learner's team interaction. If she is active in a small group, that's a good sign, and let her avoid full class discussion. You can look for other participation clues as well. Is she taking notes, watching you as you move around the room? Does she nod or make eye contact? If the answer is yes, then she is probably learning. If the learner seems completely disengaged, during a break ask if there is anything you can do to help. She will probably say everything is fine. You cannot force a person to learn, but you can do your best to engage the learner.
- One-upper. If you know you will have this person in your class, seek her out during development and ask for help. Be sure to thank her at the beginning of class for all her help. She won't one-up or criticize you if she helped you create the training. If you don't know this learner exists before you start to teach, follow the same intervention process as above.

- The leader who won't be a leader. This is my biggest peeve—when you really need support from a leader and he won't lead. I go through the same process, waiting until break time and then asking him for a moment of his time. I then ask if he can help me teach a better class. Offer suggestions on how he can use his position to help others learn. Ask this learner to speak up when he supports ideas or when the room gets stuck on an issue. Sometimes a leader thinks it's better to be quiet and let the team speak. This is great if the team is participating, but sometimes the team is waiting for the cue from the leader, but you need to let the leader know you need his help.

There's no guarantee these methods will work every time, but they usually do. Your approach must be confident when you start the tough conversations. You have worked hard to develop this class, and you cannot let anyone diminish your success. Plus, the rest of the class deserves the best, so sometimes you have to intervene. The good news is that sometimes the class culture takes over and solves the problem for you. I have had other learners lean over to the jokester or know-it-all and tell him to be quiet and let the trainer do her job. I have seen other learners conduct their own interventions during the break without me having to say a thing. As much as I like to think I am masking my feelings, I am pretty easy to read, and the learners who want to be better at their jobs don't want anyone ruining their class either.

the rule of proximity

Have you ever had a learner who chats all during class? Even if it is on the topic, it's still really hard for the rest of the class to stay engaged. This is when I use physical proximity to manage the chatter. Because I always move around the room, it's not unusual for me to stand near different learners at different times of the class. Usually, if I stand right next to the person who won't stop talking as I continue to speak, that person gets quiet right after I move there. If the chatterer still doesn't stop talking, I stop talking and stand right next to the person. The whole room usually turns to look at where I am standing, and the talker then realizes she is a disruption. I usually offer a quiet thank you to the talker as I walk away. Often, I don't even get that close to the talker—other learners tell the person to quiet down.

Measuring and Managing Engagement

The following are some clues to how you know your course is working.

- Body language. Watch for changes in body language as the day progresses. After lunch, even the most active participants' body language may wilt a little. Watch for the smiles and nods of approval or agreement, and use that to determine how many are engaged in the program. I often find myself going back to the nodding heads and smiling faces as I teach. It makes the experience easier for you if you look at your audience and focus on happy faces.
- Taking notes. If your learners are writing notes on their materials or handouts, that is always a great sign of engagement. Sometimes I like to ask learners to share what they are writing. This can act as an alternate method of teaching, allowing those learners to shine among their colleagues.
- Following you as you walk. Move around the room as you speak. Watch the learners—do they shift positions in their chairs so they can keep you in full view? This is a great tool when you are in a larger room and you want to connect with everyone, despite the fact that they are far apart.
- Asking questions. You can start with yes-no questions and then move to open-ended questions that require a more thoughtful response from the learners. Remember to tie questions to the learners' experiences. This capitalizes on ALP and helps them see the value of the training.
- Offering to answer questions others might have. I really like when one learner jumps in to help another when there's a company-specific question that I might not have the best answer for. I then add to the learners' comments with a question: Does anyone have this issue in their department? A simple conversation between two learners could help the entire class.
- Engaging in activities. I usually watch all the activities when the teams are working. I note the engagement levels of all the team members. I encourage switching which learners do the writing or the report out. This gives everyone a chance to shine and helps some of the shyer learners break out of their shells.
- Protecting and supporting their team. I love when the adult competitive nature begins to show. I have seen one learner earn a treat

but only take it if her whole team got one. I have seen teams jokingly demand more respect for their work or contend that they earned treats that should be provided to them.

- Protecting and supporting others in the class. The class culture gets even better when learners start to speak up and honor others' work. When one learner shares an idea and the rest of the class thinks it is valuable and worthwhile, the class erupts into demands that I give them a sweetie or rubber ducky. Not too many things feel as good as a room full of safety learners excited about what you are teaching them!
- Speaking to you during a break. It's always rewarding when learners who never talk in class come to you during a break with questions. The whole time you are teaching, you may be wondering if they are even listening. Then you find out they are, but they just don't like to speak in public.

Great Managers Make Great Learners and Leaders

I had the great fortune recently of having one of the best learners ever in my class. It was the CEO of an international organization. I had been training their learners all over the world. The executive team agreed that they should not only attend the training but attend with the other learners. The day came when the CEO would be attending, and I was excited to meet him. From the first moment he arrived, he showed he was ready to learn as well as serve as a leader of learning. His ease, management style, and curiosity made for an amazing experience for the rest of the class. This person did not need coaching on how to make the learning experience amazing for all there; he just knew what to do.

molding learners into leaders

If you don't think your executives or managers will show great leadership skills, work with them in advance. Coach them on how the rest of the learners will be following their lead on how much to participate and share experiences. Share with them how the new approach to training delivery includes participation and engagement, and that they can make a difference in how much the learning sticks.

Managing You

Jitters

Great design and preparation will help calm your nerves. It you are a seasoned trainer, you may be concerned only with running a different kind of program that involves more learner interaction. If you are less experienced, the best plan is to rehearse as much as you can. Get to the training location early so you don't have to worry about logistics or equipment issues. Take any extra time to relax and review your materials one last time.

System Failures

It happens to the most experienced ninja—LCD light bulbs burn out, audio doesn't work, or the dreaded automatic update to your computer cannot be stopped. If these types of system problems occur, you can start teaching your class without PowerPoint on. You should know how you want to start introductions, run the icebreaker, and energize the class without technical help. Acknowledge the issue to the class, find your stakeholder, and get assistance to fix the problem right away. This is another time when having your own instructor's manual saves the day. Referring to the manual will help keep the class going despite not having any technology.

Program Flaws

Despite all your work and a dry run, an activity is just not working. This is the time to let it go; you can fix or remove it from the next program. Recently, a client enlightened me with a phrase his CEO uses. He calls it "fail fast."

technology

As a precaution, don't kick off your programs with videos or other tech-heavy options. That way, if there is an issue, you are not stuck. Teach from the handouts and manual until the tech issue is solved. With rehearsal and a good set of materials, you should be able to teach the class without even one PowerPoint slide.

If what you are doing is just terrible, stop the exercise and move on. Don't let it drag on to meet the timing requirements. Acknowledge that this was supposed to work another way and it's just not happening. Tell your learners you don't want to waste their time and move to the next section.

Time Constraints

I am notorious for talking fast, and even faster if I am running out of time. Don't do what I do. During every break you should be monitoring your pace and ensuring you are on track to finish on time. Be ready to skip materials, shorten activities, or limit the feedback to just one team. Let the learners know what you are doing and why, keeping them informed of content changes and alerting them that they may need to skip a couple pages ahead in the manual.

Recently I had so much participation in class that I was running out of time. Lunch was ready, and I was going to the airport right after that. I indicated to the class that we would need to skip the last two role plays so they could eat and start back on time with their next speaker. There was almost a revolt! They insisted I let everyone complete the role play, and they would be happy to shorten the lunch break to allow for it. It was a pleasant affirmation that the content and materials were what they needed to learn. We finished the materials, and I left for the airport with the great feeling that I helped the learners bridge the gap.

Communication Styles

Learn what your training style is. From a young age I have enjoyed talking. I took drama in school and performed in some local plays. I have always felt comfortable in front of a crowd, but that may not be your style. You may need to work on how you can deliver your best ninja training. Some of you may be great at standing at a podium but struggle with managing the crowd during activities. Reflect on your strengths and weaknesses and look to trusted colleagues for honest feedback. Be ready to hear the truth and use it to improve your skills.

Leave Your Baggage at the Door

A question that is often asked when I teach a ninja training class is, "What do I do about the guy I know is going to be a problem or that I had a heated discussion with about some important safety topic last week?" The answer is simple, yet difficult: get over it. Don't let someone else run your class. Don't let past experience dictate the future. If you know the person is always difficult, spend time with him a few days before class. Let him know you need his help to make the class great for everyone. Acknowledge that you may have had some difficulty with him in the past, but you want to keep it in the past. Maybe that person wants to let the baggage go as much as you do. Use the opportunity to act as a leader and forget about old issues.

Film Yourself

I am sure you have heard this advice before, but filming yourself while you teach is a great way to improve. Whether you learn that you say "um" too much or that you jingle your keys in your pocket, every trainer can benefit from some scrutiny. Acknowledge that you are probably your worst critic, so don't be too hard on yourself. Choose one or two ways to be better during your delivery and focus on them.

Find a Mentor

Is there someone in your organization that inspires you? Watch how they teach and determine what makes them inspiring? Ask them to coach you on how to improve delivering your materials.

Show Your Passion

If you love to train, let it shine through! Helping people learn to be safer at their jobs can be one of the most satisfying parts of the EHS profession. I have spent most of my career honing the art and science of safety training, and it is what I love to do most. The feedback from learners fuels my future learning and instructing efforts. I might not always know that I helped prevent an accident or save a life, but I believe great training is one of the best ways to show the value and benefit of safety to everyone we teach.

Stay Healthy

Keeping your health and strength up while you deliver your materials is critical. If you run out of steam, the class may falter. I have been sick a few times when I had to teach, and usually the class is empathetic and helpful. I tell them to keep their physical distance, and I use a lot of hand sanitizer.

Because I fly to teach my classes, I can't just say let's do it next week. Honestly, you can't usually do that either. Your schedule, the learners' schedules, production, and training-room usage usually dictate that the class must go on. Do your very best and rely on the hard work you did during preparation. This is a time when activities can really save you. You can slow down a bit since you are not front and center, the learners are.

When I travel, especially on long trips teaching multiple classes, I have a daily regimen of vitamins I take. I don't know if it makes a difference, but I have always managed to stay pretty healthy, even when I am teaching ten out of fourteen days and flying to another city or country almost every night. Another healthy practice of mine is getting enough sleep. As I have aged, it has been easier to make the right choice. I have been in foreign countries when dinner doesn't start until 10 p.m. I am usually in bed by then, resting for the next day of teaching. My clients know I need sleep and proper nutrition, so they are always thoughtful about getting dinner over earlier. If I have to skip a social event to sleep, then I let my hosts know and excuse myself early. Despite a few strong customs in some cultures, I have never been in a situation where the client was angry or offended with my required sleep schedule.

Don't Leave Home Without These

I have been training and traveling most of my career, and there have been a few times when I was missing critical items to teach effectively. As a result, I carry a lot in my briefcase. Here is a list of things I'm never without.

- laptop and power cord
- adapter to fit any LCD projector ever manufactured
- electricity adaptor for international work
- USB cords to charge any equipment

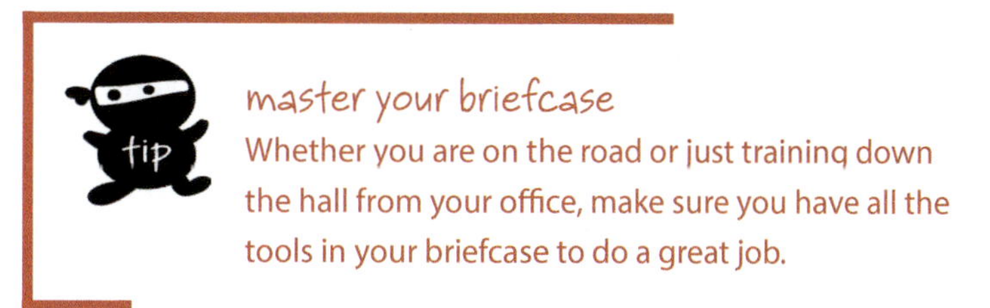

- mouse
- high-capacity jump drive to transfer my presentation to someone's computer if mine fails
- remote pointer/remote slide advancer and spare battery
- mini-speaker with extension loud enough for a room of fifty people
- hardcopy of my training program, materials, and trainer guide
- pens, highlighters, and sticky note pad
- lots of colorful markers
- business cards
- candy or sweeties, treats, and a bag to put them in
- umbrella or raincoat
- spare set of reading glasses
- hand lotion
- hand sanitizer
- over-the-counter medicine for headache, allergies, sinus pain, and tummy trouble
- tissues
- lip balm
- throat lozenges/mints/hard candy
- packets of tea and honey in case I get a sore throat
- toothbrush and floss

This may sound like a lot, but I have needed every one of those items more than a few times. A lot of the courses I develop and deliver last from one to three days, so having all I need keeps the stress level down. I cover all my electronics and peripherals in brightly colored tape so I can see them clearly and am sure to take them when I leave class. I've also invested in an excellent rolling briefcase to protect my shoulders and back.

Finish Like a Ninja

As you begin to close out your training program, make sure you have enough time to do so smoothly. Rushing tests, evaluations, and commitment statements negates all the hard work you have done. This is where your design and development need to be accurate. If you have thirty learners in the room and plan to have all of them read their commitment statements, that amounts to at least forty-five minutes of form completion and learner discussion. A posttest may take ten to thirty minutes, depending on content and complexity. You need time to go over the answers and ensure everyone has corrected their tests to 100 percent. The evaluations could be passed out before the final break. That way, the learners could complete them after they return from break and during your final instruction time.

Finally, recap the big issues you taught and give them a gracious thank you for their time. Please and thank you should be common phrases in any ninja's class.

Dealing with the Unexpected

I get asked a lot of questions about how to deal with failures, brain-dead moments, and difficult learners. Here are some ninja tricks for recovering from unexpected challenges.

I Forgot What I Was Saying

This is a great time to sneak in an engagement check. If I get distracted by a question or a learner sharing an opinion, I don't say I forgot what I was talking about. Instead, I am stealthy like a ninja and ask, "OK, so who was paying attention? What was I just talking about?" Someone always knows and then I am right back on track without anyone being the wiser.

I Get a Question I Can't Answer

I turn to the class and ask who would like to answer that question. It often starts an excellent discussion by the learners, and I get to learn too. You can always rely on the tried-and-true option: Tell them you don't know but you will find out. Just be sure you follow up and share the answer with them.

Change in Staff at Stakeholder's Office

Occasionally there will be last-minute changes to staff, either through promotions/transfers or people quitting or being terminated. One time my client was fired three days before I flew to the location to teach. I was contacted by one of the HR staff, with whom I had worked in the past, and I made a proposal on how to handle the course since the final details had not yet been worked out. She agreed to my suggestions, and the class went well. I avoided any discussion regarding the newly dismissed staff person. It was none of my business and would have only added to class distractions.

The Room Just Won't Engage

Search out the formal or informal leaders at the first break and ask for advice. In one class, I was told by a learner that their countrymen were just generally grumpy and didn't like to participate. Sometimes you have to accept a class's or country's culture and do the best you can to cover the material effectively. And remember, just because they are not asking questions and providing comments or personal stories doesn't mean they are not learning.

The Angry Learner Who Will Not Stop Ruining Your Class

Only twice in all my years of training have I had to ask a stakeholder to remove someone from the class. It sounds harsh and it was very uncomfortable, but the company had devoted a lot of time and money for the other twenty-four learners to get the information they needed. It was my duty to tell the stakeholder we were not going to get the outcomes the company wanted because one person could not stop making angry, inappropriate comments.

You Are Disregarded for Your Gender or Age

I started my work in the construction industry. I actually had someone write on an evaluation that there was nothing he could learn from a woman. I laughed at the comment. How do you overcome age or gender bias? Sometimes you won't, but if you know your materials and have followed the ADDIE process, you know you have done all you can to teach a great class. Don't let negative people impact your confidence or the class.

Some Learners Don't Know Why They Are in the Class

Sometimes managers send the wrong person to the training. If it is truly an error, seek out your stakeholder to see if the person really needs to stay or can be dismissed. Many other times learners are simply confused, and you need to use your mini-analysis time to understand why. Be sure you loop back to their concerns as often as you can. Using ALP, show them WIIFM and how the materials can be applied in their work. Keep them involved and engaged during the classwork.

Travel Problems

One time my flight was rerouted and didn't land until 2 A.M., and then I had to drive two hours to my hotel. To make matters worse, my luggage and all the training materials went to the incorrect city. That's why my briefcase is packed with so many things and is always a carry-on. The hotel was gracious and gave me a robe so I could wash the clothes I had been in all day. I drank a lot of coffee the next morning, apologized for the casual clothing, and said the class materials were expected to arrive by 11 A.M. I taught exclusively from the PowerPoint and skipped materials that were dependent on handouts. It all worked out fine, despite the inconvenience. Being honest, open, and authentic almost always works best for handling a challenging situation.

Training Culture

I once worked with a top-ten builder and had planned a half-day course. I don't remember the course at all, but I do remember the breakfast/coffee break the job supervisor provided for the training. What a spread! There was a huge cooler filled with a variety of juices, milk—and the true shocker—high-end, flavored coffee creamers. There was a toaster for the bagels and several varieties of cream cheese, as well as other pastries. The crew began eating with delight, and it was clear that this was the way safety training was done at this company. What a lovely way to show your crew you care on a pretty low budget, and what a great way to establish a positive safety training culture.

Several months later, I returned for another course and found there were no pastries, bagels, or anything else—not even coffee! This was a great group of people who wanted to learn, but they were having trouble getting past the lack of a breakfast. And then something bad happened—the lunch came. It was awful: a tiny sandwich and a bag of chips. Learners were expected to find their own drinks. They were very unhappy, and the evaluations reflected it. Several comments were made during the class about the company not caring about them.

Were the learners being selfish? Or maybe the company culture had built an expectation of good food that was part of their effective training experience. All the money spent on my fee, supervisors' time away from the job, travel expenses, and so on, were magnified by a learning interference about the food that resulted in almost no learning taking place. The company wasted a lot of money when they could have had a decent lunch brought in, or better yet, a bunch of pizzas. If budget constraints prevent you from providing a really good meal, then let the learners know in advance that they should bring their own coffee, and tell them in a respectful way. I now confirm with every client that there will be coffee and meals for every class I teach. If they tell me there is no budget, I pay for it myself. The $50 I may spend on coffee and doughnuts pays dividends by establishing a positive training experience that supports learning.

Z490.1 Supporting Material

Finally, Z490.1 Annex C.5 (ANSI/ASSP Z490.1-2016, 42) offers some final information on some key delivery points.

- Introduction
  - Present the overall picture of the training material and the impact it can have on everyone.
  - Stay focused on your critical training objectives.
  - Explain the program benefits and what they can expect at the end of class.
  - Explain to learners why they are being trained.
- Main Body
  - Present required and desired information.
  - Use all the tools, ALP and activities you have defined in your training plan.
- Conclusion
  - Plan and rehearse.
  - Capitalize on the laws of learning to make the final statement memorable.
  - Provide your evaluations, quizzes and commitment documents within plenty of time before the class end time.

Are you an acceptable trainer? Section 5.2.1 of Z490.1 (ANSI/ASSP Z490.1-2016, 20) indicates that every EHS trainer meet the following standards:

- understands completely the course learning objectives
- is familiar with the course materials
- is familiar with and practices using the primary and alternate delivery strategies as designed into the course materials
- in the case of a virtual environment, will have participated in adequate practice sessions to ensure a complete familiarity with

the delivery systems and will be familiar with backup plans in case of system malfunctions

Annex C.1, Training Delivery (ANSI/ASSP Z490.1-2016, 39), also references the hallmarks of effective training:

- establish a positive atmosphere or learning climate in which people can participate in a productive way;
- make participants aware that they are free to make mistakes and experiment with ideas and behaviors;
- describe their role as a guide, facilitator and trainer;
- express specifically that their priority is meeting the trainees' learning needs;
- describe and have in writing the learning objectives;
- describe and have in writing an agenda of how the training will flow and be delivered;
- use all materials and aids designed for the course;
- solicit trainee responses to keep the session active and to be able to assess learning;
- have backup and contingency plans in place if the specified delivery is not effective;
- be able to manage the physical facility so that learning is promoted.

Conclusion

You did it! You have finished your first ADDIE-based EHS training program. Congratulations! Now it's time to look at your evaluation systems and any follow up needed.

References

ANSI/ASSP Z490.1-2016. 2016. *Criteria for Accepted Practices in Safety, Health and Environmental Training*. Park Ridge, IL: American Society of Safety Professionals.

Chapter 7
Evaluate Your Program

You have made it through your first class as a ninja. Be proud of yourself! Now it's time to review the results of your program and your learners' evaluations. You can take the feedback you received and use it to improve, update, and alter your programs in the future. A good evaluation serves two purposes: it helps you reflect on what you taught as you review the reactions of your learners, and it loops back in all the hard work of creating an ADDIE-based program to bridge the learning gaps you identified way back in the analysis stage.

In this chapter, I am going to discuss the gold standard of training evaluation methodology and introduce a new (and I mean really new!) and different way to collect data on the effectiveness of your program. All the reasons to evaluate and review your program are the same, but the final tools may differ.

Why should you evaluate your program?

- To determine if your learners actually bridged that knowledge or skill gap with the training you provided.
- To create a better program. Review how the program went and determining what you can improve will make your program better.
- To help supervisors and managers determine if there is an observable change in the knowledge or skill of the learner.
- To make your stakeholders happy. Stakeholders like to know that the investment was valuable. You can gather the results of learner evaluation and feedback and provide anecdotal information through your verifications of learner engagement.

- To verify learning of regulations. Evaluations also serve to meet governmental or consensus organization requirements that you need for you files, legal purposes, or competency verifications.

There are several innovators and thought leaders in the area of training evaluation. I have tried to take their most important concepts and scale them down while providing relevant examples and job aids for the EHS industry.

Donald L. Kirkpatrick

The gold standard for evaluating training was developed by Donald L. Kirkpatrick in 1954. The concepts came from his PhD dissertation, and later they grew into *the* way to evaluate training. He used four simple words that have been standardized by the learning industry and expanded upon by his family members and other great thinkers in the ISD and business world. Kirkpatrick's four levels of evaluation focus on the major outputs you can/should measure after training (Kirkpatrick and Kirkpatrick 2016):

1. Reaction: the satisfaction the learner expressed after the training was completed
2. Learning: how well the learner acquired the knowledge or skills
3. Behavior: the extent of the learners' behavior changes after they return to work
4. Results: the positive impact the change has on the business

Let's break those concepts down a little further.

Level 1: Reaction

As ninjas, we have to look at more than just did the learners like the food, was the room cool enough, did the trainer tell great stories, and did the class finish on time? Yes, these are important because if the learners tuned out based on learning interference—especially on something you could have easily controlled—you just wasted their time and yours. Reaction to training can tell you more than just "the food was good." The trainer can gain valuable insights on the immediate thoughts and feelings of the learners. Reaction level measurements are often called "smile sheets." The smile sheet usually includes questions such as those in table 7.1.

TABLE 7.1 Standard Level 1 Evaluation Form

Please answer the following questions:	Rate from 1 to 5 (5 is best)
I was able to relate each of the learning objectives to the learning I achieved.	1 2 3 4 5
I felt that the course materials will support my success.	1 2 3 4 5
My learning was enhanced by the knowledge of the facilitator.	1 2 3 4 5
It was easy for me to get actively involved during the session.	1 2 3 4 5
I was given opportunity to practice the skills I am asked to learn.	1 2 3 4 5
I am clear about what is expected of me as a result of going through this training.	1 2 3 4 5
I anticipate that I will eventually see positive results from my efforts.	1 2 3 4 5

Ok, I will admit that this is one of my smile sheets, and it is not as bad as some. Notice how limiting the questions and potential answers really are.The sheet places a bias on positive ratings. Of course, being the designer, developer, and trainer of the course, I want good evaluations. But that is me being a business owner and wanting to have the client hire me again. If I get a lot of fives on the smile sheets, I will usually get more work, or at least a good reference. The simple Level 1 smile sheet is almost exclusively about how the learners feel right before they leave the class. But how will I convince my stakeholder that the information I received from this posttraining survey changed the learners' behavior or skill? It won't, and that was the whole purpose of the training, right? Didn't we decide that way back in the analysis phase of ADDIE?

Most EHS training evaluations stop at the Level 1 reaction phase. It is not our fault. No one told us there were other levels to consider, and certainly no one gave us great examples that we can use with our own learners. In defense of all trainers and training providers, don't judge that we are not going after the ultimate evaluation goal. A lot of companies don't want to do what it takes to really find out if the training and business goals were achieved. And most EHS professionals don't have the time to develop the tools to find out. More on that later.

Z490.1, section 6, provides information on the concepts of evaluation and some very helpful information. The reaction level information in 6.2.1 refers to the information as subjective (ANSI/ASSP Z490.1-2016), which is exactly right. Learners are revealing what they feel after the class. That is valuable, but we need to know more.

There may be a few seasoned ninjas out there who have been using more complex evaluations from Level 2 and beyond, but I haven't met many. If you are out there, share your samples with other EHS trainers. It would be great if we could help the rest of our profession.

Level 2: Learning

Recall that this stage is about determining if the learner gained knowledge from the materials presented. The EHS trainer default is to administer tests to find this out. Z490.1, 6.2.2 (ANSI/ASSP Z490.1-2016, 25), refers to a variety of different testing tools, including:

- written test
- oral exam
- completing a project based on the content delivered
- demonstrating the new, correct skill or knowledge within the learning experience or via a simulation

I mentioned in a previous chapter that I don't like the word *test*. I know it is all about the semantics, but tests are for children, and using that word could negate all the hard work you did when you implemented ALP. Consider using a different term, such as evaluation, skill confirmation, or whatever you think works with your learners. But if your learners don't mind the word *test*, then forget what I just suggested.

Other ways to confirm learning can be more informal but may not yield an official piece of paper for your files. Consider informal OJT evaluations, follow-up conversations, toolbox talks, or morning/shift-change huddles. The supervisor can talk about what was covered in class and use the time with teams to discuss how the new skills are working and if there is anything hindering their attempts to implement what they learned.

If after a formal or informal evaluation you are not seeing the direct result you are hoping for, it is time to determine why. Is the training you delivered at fault, or is the employee missing the skills sets or equipment to do the job safely? This is where job aids for supervisors and your fellow EHS professionals can help. Ultimate success will have to be measured over time, not just on the day of the course, but Level 2 learning needs to be measured shortly after training since we are trying to confirm the participants learned what you wanted them to learn. You may already be doing this type of observation as part of the rest of your EHS duties. Two days after training is delivered, you may head to the shop floor to confirm no one is bypassing the newly installed guards discussed in your class, or perhaps the day after class all the supervisors are using a form you gave them to confirm all ladders are inspected before use. Consider the value of linking the onsite evaluations occurring at that moment to your training outputs. Instead of conducting a compliance survey of a new policy, you are looking at whether the learning stuck. Let's be honest, evaluating whether learning sticks sounds a lot better than doing a compliance audit. The above options can lead us to the next level of evaluation. And this one gets more complicated.

Level 3: Behavior

Let's assume you confirmed the class full of learners has demonstrated they can use the new fall protection equipment you purchased and trained them to wear. You, a supervisor, or someone competent has watched each learner properly inspect, wear, and store the new equipment. That is great news, but do you need to know more? Do you need to know they will wear this equipment and use it the right way and that they are motivated to wear it every time, all the time, correctly? If yes, you need a Level 3 behavior evaluation. Now that we are talking about motivation to implement the new knowledge or skill, we see how it's gotten more complicated. Maybe this is why we don't see this level of evaluation across the EHS industry, at least not formally.

Z490.1 supports these ideas in section 6.2.3 (ANSI/ASSP Z490.1-2016, 25). It reminds us that when the observation of performance reveals that we missed bridging the training gap, we have to go back and find out why. The reason the gap was not bridged could include problems with the training design or delivery. This level of evaluation could also highlight nontraining issues that prevent implementation of the new behavior or skill, such as a lack of the right equipment on the job, new employees who haven't been trained yet, or one of the worst options, information from a supervisor or others that doesn't support the new behavior or skill. As EHS professionals, we see this more often than we would like to. The mixed messages sent to learners imply that while safety and the training are important, don't forget that all the production deadlines must also be met.

Level 4: Results

This is the least evaluated stage, and for good reason. It can take more time and resources and can be hard to prove your training is the cause of positive safety results. Let's be honest: If safe behaviors of any kind are improving, people and other programs you are implementing will battle for the credit for this upturn in safety. How great would it be if the training got the credit for changing behavior or skills? More often, an audit or inspection counts the number of noncompliant behaviors, catching your learners doing it wrong and punishing them. What if we made it about confirming the desired outcome of the training and crediting the results to the training? Think carrot instead of stick.

By first using the Level 3 behavior observation method within a few days or weeks of the training program to confirm behavior, you can now link that measured and documented behavior to your end goal: successful results. It might be a long time out, but can you connect the training topic to the following things?

- decreased accidents and illnesses concerning the training topics
- higher audit scores when conducted internally or by a third party
- increased favorable observations of the correct behaviors
- decreased near miss observations or submissions
- positive feedback from learners
- increased safe behaviors connected to production value
- lowered insurance costs due to decreased severity or frequency

Z490.1, section 6.2.4, also refers to the need to connect organization results and to link overall organization performance. One way to do this is to isolate one group from the training and compare behaviors, lagging indicators, and other business objectives. This can be tricky for EHS training since we can't always wait to train everyone because of a regulatory deadline or some life-critical knowledge or behavior that is needed. But if it is possible and safe, give it a try. Some key safety measures Z490.1 includes are:

- increased learner safe behaviors
- increased learner training implementation
- reduction in accidents, near misses, and illnesses
- reduced work-related insurance claims (ANSI/ASSP Z490.1-2016, 26)

review your evaluation forms

Talk to HR or your learning department and find out if you must use a standard corporate evaluation. Find out if you can edit it for the EHS training, or better yet, make significant changes based on what you have learned. If they are open to the change, ask them to help. Their feedback can be valuable. Who knows, maybe they are ready for a change too.

Jack Phillips

If you didn't think that Levels 2 through 4 were hard enough, Jack Phillips adds to the levels by promoting the need to determine the return on investment (ROI) for training programs (Phillips and Stone 2002). I am only going to touch on his concepts because, although they make great sense, even full-time training professionals don't attempt to derive ROI on more than a small percentage of *all* the programs they oversee, with EHS programs being a small portion of that small portion.

Isolating the value of EHS training is almost impossible. Proving that an accident didn't occur because a learner implemented what he or she learned is even harder. To pinpoint an exact return on the organizational investment is a high standard to reach, but that doesn't mean you should not employ thoughtful resource management and attempt to tie changed behaviors or knowledge to your work. You should still keep track of all costs to implement the training program and any follow-up costs as part of the verification of learning.

Notice how I am trying to draw a relationship to what you will need to know about the learners and what you may already be doing as part of your daily EHS responsibilities. By making it part of the full ADDIE training cycle, it shifts the "why" of an audit to a confirmation of learning. The key is to develop job aids for you or anyone else who may need them to confirm that behavior has changed, if it will stay changed, and that you can track it back to the learning and business objectives you worked so hard to achieve.

Will Thalheimer

It wasn't until I was reading a book by my favorite training mentor, Elaine Biech, *The Art and Science of Training* (2017), that I first learned of Will Thalheimer. Will's book, *Performance-Focused Smile Sheets*, is exactly as he claims, a "radical rethinking of a dangerous art form" (Thalheimer 2016). I have been a follower of Kirkpatrick's four levels since I began learning about ISD, but Thalheimer's approach to an expanded, deeper version of a smile sheet resonates with me, primarily because it absorbs some of the best elements of Levels 2, 3, and 4 and incorporates them into the already accepted concept of a smile sheet. I see it as an easier alternative to mastering Kirkpatrick's levels while still honoring the time-proven system. I like things as simple as possible, and I think Thalheimer has given us an opportunity to achieve a smile sheet with more value and, hopefully, that requires less work. Note that I didn't say better Level 1 evaluations because this is about incorporating some of the greater goals of evaluation into the smile sheet.

Thalheimer believes that there are four sets of guidelines for creating better smile sheets.

1. Develop a better quality of output from the learners' decision-making when answering evaluation questions. Keep these recommendations in mind when designing your new smile sheets:

    - Remind them of what they learned before you ask the questions.
    - Keep the evaluation to only the information you need.
    - Make sure they have the right amount of time to complete the evaluation.
    - Avoid leading the learner with biased questions, such as "Was the trainer well qualified?"
    - Keep the questions clear and relevant, Thalheimer suggests avoiding questions about the food or room temperature, because if the learners are unhappy, they will share those feelings in the comment section.
    - Only ask questions the learners can answer with true knowledge, avoid asking questions that they subjectively answer, such as "My learning increased in this class."
    - Ensure the answer choices you give for your questions are descriptive and clear.

- Entertain the idea of doing a follow-up smile sheet a few weeks after the training.

2. Ensure the questions clearly rate the different levels of success for learning. This means avoiding the much-used but of-little-help Likert Scale (that "rate your opinion on a scale of 1 to 5" option) or subjective words such as "strongly agree" or "neutral." It is very difficult to attach an actual measurement to undefined numbers or words. For example, the difference between "agree" and "strongly agree" can vary quite a bit from one learner to the next. Some people think a rating at the top of a Likert Scale, in our example a 5, is reserved for a training class close to absolute perfection, while others might rate a 5 for a class they thought was "pretty good for a safety class." Define the words you will use to rate the questions and establish a measurement before you distribute the evaluation. For example, consider this evaluation question: "Since I have completed this chemical hazards training course, I have adequate familiarity with the requirements to keep me safe." Is adequate familiarity the highest standard of knowledge you want to achieve, or is it mid-level? What are the other key phrases you can use to determine how well your learners can use what they learned. Don't worry, I have some examples for you.
3. Measure only the things that really matter. For the EHS professional, this usually means, do they remember what they learned, will they apply it, and are there systems in place to keep the behavior or skill active?
4. Use the outputs to enlighten stakeholders. Finally, how can we use this information to justify the time and money spent on great training and, more important, turn our stakeholders into advocates of great training. Thalheimer believes that the old way of using charts and graphs that show how many people were trained and that the trainer received a score of 4.4 doesn't resonate with the upper-level stakeholders or C-suite professionals. Instead, use the information about changed behaviors as a result of training derived from your "radically rethought" smile sheets. Provide insights to stakeholders based on the measurements of learning stickiness. Using your version of a job performance checklist in posttraining can help you see

where the training is working, so provide that information to C-suite.

The results of these types of questions can also point you in the direction of improvements in your training program. If everyone answers that they cannot perform the skills or don't have the knowledge you were trying to teach, then maybe the materials or the instruction needs to be improved to ensure the learning objectives are met.

Tables 7.2, 7.3, and 7.4 list some sample questions that you can adjust to fit your topic and measurement scales.

Do you see what is happening here? We have developed our follow-up action to the evaluation as we have written the evaluation. We may even see opportunities to develop new job aids based on the feedback. That's amazing, and something I haven't heard of in EHS training (if you are doing this already, please share the info). Now imagine you did all of this during the first stages of the ADDIE cycle. You are looking at powerful evaluation tools that you can analyze and that can provide practical data to your stakeholders!

Is there more to learn about Thalheimer's approach? A lot more, and this ninja plans on reading his book several more times. But right now, I have to go and rewrite every smile sheet I have ever done because they are sorely lacking!

Table 7.2 Regarding Training Effectiveness

Based on the material you learned today, how able are you to complete the daily fit test and clean and store your respirator. Please choose one.

- o I will not be able to fit test, clean, and store my respirator without help.
- o I have general awareness of how to fit test, clean, and store my respirator but will need more practice to get it right every time.
- o I am able to fit test, clean, and store my respirator but will need help remembering the steps after tomorrow.
- o I am confident I can fit test, clean, and store my respirator after every time I wear it.

Answer choice	**Standard***
Will not be able	Unacceptable
General awareness	Acceptable with OTJ follow up from supervisor or EHS staff
Am able . . . but need help remembering	Acceptable with a job aid
Confident . . . every time	Awesome

* Although it may seem obvious to you what the answer choices mean, a clear standard should be developed in advance so there is no confusion during the analysis of the data.

Table 7.3 Regarding Learner Motivation

After today's presentation, how motivated are you to adopt the new driving procedures when you get back to the job tomorrow? Please choose one. o I will not make the new procedures a priority. o I will make the new procedures a low priority. o I will make the new procedures a medium priority. o I will make the new procedures a high priority. o I will make the new procedures my highest priority.	
Answer choice	**Standard**
Not a priority	Unacceptable
Low priority	Unacceptable
Medium priority	Acceptable with OTJ follow up from supervisor or EHS staff
High priority	Acceptable
Highest priority	Awesome

Table 7.4 Regarding Implementation

Since you have completed this course on supervisor leadership skills, how well can you implement the new opportunities to manage presented in the materials? Please choose one. o I have significant weakness in my abilities to implement the concepts. o I am familiar with the concepts of safety leadership. o I have a solid understanding of what I need to implement with my team. o I am ready to begin implementing the concepts with my team.	
Answer choice	**Standard**
Significant weakness in my abilities	Unacceptable
Familiar with the concepts	Unacceptable, requires OTJ follow up from manager or EHS staff
Solid understanding	Acceptable but requires observation of implementation by manager or EHS team
Ready to begin implementing	Awesome

Figure 7.1 is a job aid for helping supervisors or EHS professionals observe learners to see if the desired behavior change is happening. Notice how the design allows for learner training retention evaluation and follow-up feedback and action opportunities to assist learners who didn't retain all the information from the training.

Training effectiveness observation tool

Observer name	REGINA MCMICHAEL	Today's date	12/1/2017
Observee name	KELLEY SPENCER	Training delivery date	11/28/2017

Task observed	Equipment, tools, or process used
PAINT APPLICATION AREA IN CELL B, BUILDING 417	FULL FACE RESPIRATOR, BRAND XX MODEL XX SPRAY SOCK

Instructions

*Introduce yourself to individual (if needed) and explain you are doing follow up on the recent training they completed. Ask the individual to complete the task you wish to observe. Conversely you can observe the induvial causally while they are spontaneously performing the task in review.

Evaluation criteria (check the appropriate box)

	Individual is not be able to fit test, clean and store respirator without help.
	Individual has general awareness of how fit test, clean, and store respirator but will need more practice to get it right every time.
X	Individual is able to fit test, clean and store respirator but needs help remembering the steps until proficient.
	Individual demonstrates correct procedures for fit testing, cleaning and storing of respirator during every use.

General observations and notes

KELLEY WAS ABLE TO DO THE PROCESS PROPERLY BUT SINCE SHE ONLY DOES THIS PROCESS A FEW TIMES A MONTH SHE BELIVES SHE WILL STRUGGLED TO REMEMBER ALL THE STEPS SINCE SHE IS NEW TO THE PROCESS AND WORKCELL.

Follow up recommendations

PROVIDE KELLEY THE JOB-AID ON RESPIRATOR USAGE AND HAVE HER STORE IT WITH HER RESPIRATOR SCHEDULE A FOLLOW UP WHEN SHE IS NEXT SCHEDULED TO USE THE RESPIRATOR AND ASSIST IN PROPER USE.

Observer signature *Regina McMichael*

* Be sure you explain that you are conducting training follow up to help use their new skills or knowledge. Put indivual at ease to ensure you are getting an accurate undestanding of the new skills or knowledge.

FIGURE 7.1 Training effectiveness observation tool

Managing the Review of Your Training

How do you deal with a poor review, or worse, a bad review? The first thing to remember is: keep the comment in context. Did several learners comment in a negative manner? Do other learners provide positive feedback on the same program? I was coached a long time ago to ignore the best and worst reviews and then look at what is left. Regardless of what type of evaluation forms you are using, always leave space for comments at the bottom. There is always something we can improve.

Were the Comments Constructive?

Recently, I received some feedback from a conference session where an attendee said he learned nothing new. Although I was disappointed there wasn't even one nugget of learning to be found, I let that comment go. I completed the abstract of the program as accurately as I could and offered the learning objectives at the beginning. Since I could not do an analysis on an unknown conference attendee, I just had to accept that maybe this person should have attended another session. If the comments are constructive, with comments such as, "Job aids supporting the concepts of the learning objectives would have been helpful," then that is on me. And you know what? After researching and writing this book, I am going to take more aggressive steps to include more job aids in all my programs. Other constructive comments that can help you improve as a trainer include recommendations on the following:

- pace of the program
- quality of your presentation materials
- interaction with learners
- A/V value or quality
- technical knowledge in the subject
- accuracy of the training description in advertising materials versus what was taught
- opportunity to practice skills in the class
- demeanor of facilitator
- application of ALP

- the fun factor
- the dreaded lecture factor
- appropriateness of the materials for the audience
- likelihood of applying knowledge or skill to work

Were the Comments of a Personal Nature?

I am sad to say that far too many times in my career I have received comments about the effectiveness of a woman conducting technical training. I can recall comments regarding my age or experience as well. I fear there may be other comments based on other issues that are irrelevant to the qualifications of the trainer. I just laugh those off, however. If the learner is someone you interact with regularly, then you may need to seek out your stakeholder for support. If this learner's bias prevents him from learning, then something needs to change (and I don't mean you). Comments that are personal can hurt, and even if they are not true, many of us will let that one negative statement overshadow all the positive comments. Don't let anyone do that to you. I recently got a comment that I was "trying too hard." How can anyone try too hard when teaching is their passion? If you did your job and taught with passion, compassion, and knowledge, don't let their problem become yours.

have someone else read your reviews

I recently learned this great trick: Don't read your program evaluations; have someone else do it. Ask them to share all relevant information to improve your materials, but ask them to not share personal attacks or comments that won't improve you or the program. I think this is brilliant! If one person says one negative comment, it can impact me for a long time—and maybe not for the good. If there are a series of comments about my style or content, then, yes, I need to know about that. But if only one person out of fifty thought I was too upbeat, I probably won't change my style and reading that comment won't make me a better ninja.

Were the Comments about Nontraining-Related Issues?

Although I am sure you would prefer to have comments limited to the work you did, sometimes you can get insights into the culture of the organization or discover issues that interfered with the opportunity to learn. It can be helpful to know these types of things in advance, and sometimes you can discover them in the analysis phase of ADDIE. But if you didn't, use the new information to help with future programs. If the comments warrant extra attention, speak to the stakeholder about them and determine if there is anything that can be done that will improve future learning.

Were All the Reviews Amazing?

I include this one because, just like we need to toss the biased, negative comments, we need to temper our response to positive comments. OK, so you had a great class, you were a ninja in every way possible, and I am thrilled you got feedback that makes you feel confident. Just don't use that as an excuse not to work as hard for your next training program. Remember the law of recency: The next set of evaluations will be about the class they most recently attended. So, work hard to make every class exceptional. Table 7.5 can help you do that. It presents an example of a training progam self-audit.

Overwhelming Culture

I was teaching outside the United States in a location where, no matter which ninja trick I used, the learners were just not engaged. They had tuned me and the materials out. It's possible they didn't want to hear it, there was learning interference that made my words appear empty, or they just thought my program was lousy. It is hard not to take it personally when a class is sending negative feedback through lack of participation or body language. One of the few active participants came to me during the break. He smiled at me kindly, and said, "Some people have no ears." It took me a moment to understand what he meant.

As trainers, we just have to accept that there is more going on in a class or corporate culture than we can change, and there are some learners that don't want to learn, won't participate, and don't want to change their knowledge or skills. It didn't matter what I said—they did not have the ears to hear me. His insights helped me focus on who wanted to learn instead of over-focusing on those who didn't.

TABLE 7.5 Training Program Self-Audit

Training Program Self-Audit	
Program name:	
Training date: **Review date:**	
Before training	
Course material:	**Completed?**
• Reviewed existing training materials and revised them as required.	
• Used learner-focused design strategies for learning retention.	
• Confirmed the ADDIE model was followed.	
• Adjusted training program during final review/after dry run	
Audience:	
• Confirmed learners attending are the correct audience.	
• Divided learners into separate classes or working groups as needed.	
• Identified any stakeholders, managers, or supervisors who should attend, support, or be knowledgeable about the training.	
Training location:	
• Confirmed training location, agenda, food and beverage, and A/V equipment.	
• Confirmed training location was safe in which to learn.	
• Provided training location to all learners.	
During training	
• Established yourself as a credible instructor.	
• Introduced supporting subject matter experts.	
• Presented the agenda, learning objectives, and expectations.	
• Modeled professional behavior.	
• Demonstrated flexibility in training delivery as needed for learners.	
Adapted to the learning styles of the audience:	
• Utilized adult learning principles.	
• Encouraged active participation in discussions, working-group activities, and Q/A opportunities.	
• Maximized presentation design and delivery to achieve objectives.	
• Adjusted activities to reflect needs of learners.	
Used instructional methods appropriately:	
• Balanced use of handouts, job aids, videos, etc.	
• Used 6x6 rule for presentations (six words per line and six bullets per slide).	
• Implemented blended training (when applicable).	

TABLE 7.5 Training Program Self-Audit (cont.)

Maintained effective communication skills:	**Completed?**
• Demonstrated subject expertise.	
• Facilitated with learner-appropriate language, tone, technical jargon, and voice volume.	
• Avoided slang, inappropriate topics, and cultural messages.	
• Made eye contact and encouraged participation verbally.	
Demonstrated dynamic presentation skills:	
• Used gestures, silence, movement, posture, space, and props effectively.	
• Shared anecdotes, stories, analogies, and humor effectively.	
• Demonstrated authenticity with words and body language.	
Facilitated through questioning and obtaining feedback:	
• Created a list of questions to be answered later as needed.	
• Responded positively and appropriately to questions.	
• Used active listening techniques.	
• Repeated, rephrased, or restructured questions as needed for other learners.	
Provided reinforcement:	
• Reinforced desirable behaviors.	
• Politely corrected unfavorable behaviors.	
• Encouraged those who don't participate.	
• Balanced feedback coming from different participants.	
After training	
Evaluated training:	
• Considered evaluations beyond basic smile sheets.	
• Administered assessments (Level 1, 2, 3, or 4).	
• Reviewed assessments and adjusted future training and follow-up activities.	
• Reported assessment to the stakeholders.	
• Recorded attendance of learners, add to LMS as available.	
Notes:	

You may have noticed that this chapter is shorter than some of the other elements of ADDIE. I did that on purpose. You have a lot of new information to process, and I believe the focus of your efforts should be on the "DDI" phases of ADDIE. Analyze and Evaluate are also very important, and I want you to know I respect the constraints most EHS professionals face finding the time to adopt all these new ideas. If the masses scream for it, I can write more on A and E in the next book.

References

ANSI/ASSP Z490.1-2016. 2016. *Criteria for Accepted Practices in Safety, Health and Environmental Training.* Park Ridge, IL: American Society of Safety Professionals.

Biech, Elaine. 2017. *The Art and Science of Training.* Alexandria, VA: Association for Training Development.

Kirkpatrick, James D., and Wendy Kayser Kirkpatrick. 2016. *Kirkpatrick's Four Levels of Training Evaluation.* Alexandria, VA: Association for Talent Development.

Phillips, Jack J., and Ron Drew Stone. 2002. *How to Measure Training Results: A Practical Guide to Tracking the Six Key Indicators.* New York: McGraw-Hill.

Thalheimer, Will. 2016. *Performance-Focused Smile Sheets: A Radical Rethinking of a Dangerous Art Form.* Somerville, MA: Work-Learning Press.

Chapter 8
Invite Others to Join the Ninja Team

We know we can't develop and deliver a new training program alone. A good ninja will leverage any and all resources available to deliver a great training product. Whether it comes from inside the company you work for, the client that has hired you, or someone who side hustles, people are willing to help. Not everyone understands the ISD model and why the system should be followed, but with some explanation and your passion to create great programs, you might be able to convince them to work with it.

Your Stakeholders

Without the support of your stakeholders, your projects will always be harder than necessary. This is the first group of people you should bring on your team. You will need to take the time to explain the ADDIE process and why it will benefit them. Share with them the other organizations that use ADDIE, talk about the ANSI/ASSP Z490.1 standard and the value of using established methodology. Make sure they understand that initially the process may take more time but that it will give you the tools to measure the learning and business objectives. Tying the business plan to the process will help the financial staff and the nonsafety and nontraining teams understand the "what" and "why" of what you are doing. Use this book to remind yourself of all the tactics and skills you have learned to develop your persuasion plan.

Marketing, Brand, or Communications Department

If you have a department with a name similar to one of those listed above, its staff may have some rules you need to follow when you develop learning materials. Connect with them at the analysis stage of the project. There's no point designing materials they won't let you use. You could try to evade the rules by doing what you want anyway, but that puts you, your learners, and the manager at risk. Plus, you certainly won't look like a pro. Ask them about the following:

- required PowerPoint templates
- allowable company fonts
- company color scheme with the RGB numbers (more on this in chapter 10)
- art, photo, and cartoon usage policies
- video production guidelines
- online posting guidelines
- social media guidelines
- legal prohibitions for showing proprietary or military/governmental contract-protected images
- privacy policies
- suggested image vendors
- use of the company video camera (and camera operator, if you have one)
- editing standards (ask if they can help in that area)
- copyright use policy
- content permissions policy
- anything else they think you should know

I know this may seem extreme, but I have experienced issues with every one of these topics as an employee or consultant. I've seen programs that didn't get printed because copyright usage permissions were missing. I've seen trainers delivering programs when the VP of marketing walks by the classroom and sees clip art on the screen and goes bananas. I've seen a US$20,000 invoice from a stock photo provider for unauthorized use of a copyrighted image. I've seen film of top-secret processes make its way into training programs.

View this department as your guardian angel, not as your enemy or a hindrance to completing your project.

Subject Matter Experts

Your subject matter experts (SMEs) can become your biggest advocates or your toughest opponents. I have experienced both. I have worked with a lot of engineers in my career, and some of them love the ADDIE system. Why? Because it is a system! Conversely, some SMEs don't like change. Heck, lots of people don't like change. So, just as you have to persuade your learners of the value of adopting new knowledge or skills, you will likely have to do the same with some of your SMEs.

Who is likely to be your SME? It could be a technical staff person you need to break down a process during the analysis phase so you can fully understand what the learning gap is. It could be an engineer designing a new process that has elements of EHS that you need to teach your learners. It could be a vendor who is selling you new equipment but doesn't plan on helping with the EHS element of the instruction. It could also be allied EHS professionals in your organization who may not conduct training but will be a big part of content development.

To avoid receiving a brain dump of their knowledge—or worse, thirty PowerPoint slides with full pages of information they expect you to teach—take the time to share ADDIE with them *before* they give you materials. If you try afterward, they will show more resistance since it means more work for them. Here are some ideas to consider as you bring your SMEs into a project:

- Remember, they already have full-time jobs, and they likely love the content they create. It is their expertise and maybe their passion.
- Invite them into the ADDIE process. It's likely they can provide insights in every phase of the cycle, from doing a needs analysis, to formulating the right evaluation questions, to designing job aids for posttraining support. Also ask them who else might be able to help on the project. Perhaps they know a learner who can help you with the ADDIE process based on their product or process experience.

- Acknowledge they are the experts, but gently remind them you are running the project. You do not need to be the ultimate expert on the topic to develop great training materials, especially if your SMEs will be part of the delivery.
- Clearly define the roles and responsibilities of the SMEs and on what part of the project each will be working. If you haven't written your learning objectives yet, you may want to bring them on board for that, but only after you help them understand the process of learning-objective creation. Confirm with them that not only do the learning objectives achieve the goals of the course, but that supporting content should only include what is needed to achieve the goals. Some SMEs are not creative types, so moving them away from slides filled with text and a lecture format must be handled respectfully. Explain that some of the learners will change behaviors or gain knowledge faster with activities and group learning instead of through a traditional lecture.
- Set dates for deliverables to match your development schedule. If they are busy and this project is not a priority, you may need to remind them regularly of deadlines and encourage them to meet the dates. Leveraging your stakeholders in the process may help the SMEs free up the needed time to help.
- Show them relevant examples of PowerPoints, scripts, or job aids that may help them visualize the end goals of the program development. If you are dealing with very technical people, examples can be very useful.
- Create forms, tables, or questions to answer. This allows the SMEs to give you the information in the format you need, and it helps them stick to just what you want them to tell you.
- Put them in the learners' shoes. If they are overwhelming you and the project with information, ask them to imagine how the learners will be successful in the training with the time available.
- Build a positive relationship. They are experts, right? Respect what they know and can offer your program development while gently keeping them inside the ADDIE cycle.
- Take the time to read what they give you. Even if it is a lot of information, they deserve to know you reviewed their work. Besides,

how else will you get them to focus on necessary content if you don't know what they provided to you?

- Acknowledge that the materials they provide may be previously developed work for another project. In this instance, your critique of the content could get personal. Try to remember what it was like the last time you received a critique of one of your projects. Use what you learned from that to be a great partner.
- Be responsive to their needs. Plotting out a schedule to work around their deadlines and managers' requirements goes a long way in maintaining a good relationship.
- Try to learn their language. If they are experts in maintenance and you need them to develop the lockout/tagout training, then take the time to become familiar with their terms and processes. You might need to know all of that to deliver the course, so it may supply double benefits.
- Be appreciative. Even if they are forced to be on your team, thank them for their help. Keep their manager in the communication loop so you can share the value they bring to the project. This is good advice for any project or program you work on.
- My favorite ninja trick is from ISD professional Claudia Escribano: "When all else fails, be prepared to make things up. Yes, I know that may sound heretical, but it works because SMEs can't stand to see their content portrayed incorrectly. So, if you're not getting the help you need, do the best you can with what you do know, and ask for feedback. Don't worry that it may all be wrong. Once the SMEs see that, they'll realize how much you need them, and they'll jump in to make sure that the content is accurate" (Escribano 2009).

After sharing the ADDIE system and using the tips above, you may not just get some buy-in, but you may find an SME to champion your cause. While employed by a large insurance company, I worked with a lot of SMEs, many of whom were engineers of every type and variety. One engineer was creating a training program for his specialty group and his customer base. Because it was customer-based, he was encouraged to consult with me for assistance. I explained the ADDIE cycle, what learning objectives were supposed to be and why, and then brought in some ALP and activities.

find the right expert

Seek out SMEs for your team who might be more open to ADDIE and ISD. They can be invaluable in convincing other SMEs that the process is worthwhile, and they may be able to highlight specific benefits the system provides to the company.

Not only did he get it, he went back and edited existing programs based on his newly learned ISD techniques. He was a huge supporter of the work I did, and he acted as a champion when the other SMEs were skeptical of ADDIE.

Working with Your IT Group

If you work for an organization that limits software you can use, downloads you can initiate, or websites you can visit, then you are in good company with many of your fellow EHS and ISD professionals. Almost everyone who works with a computer gets discouraged by the IT group's limitations. They say it's for the good of the company—and it probably is—but that doesn't mean you can't get them to lift or alter some rules for you. You just have to prove it won't hurt the company.

But first, you have to get them to talk to you. If you work for a larger corporation, even finding the right people can be difficult. I always start with any available helpdesk you may have to call if you have computer trouble. If you know where they sit, head over there and see if you can make friends. They may be wary of your new friendship; you are probably not the first person who tried to get them to change their rules. After establishing a positive relationship, find out whom you can talk to about trying some thirty-day software trials on products you want to learn more about or experiment with. Sometimes all you are asking for is access to cloud storage for huge PowerPoint or video files, or it may be something more interesting like an e-learning development software package. Regardless, find that person or department and let them know what you are doing and ask for help. Don't try to skirt the system at work. If you get caught, you could lose access to that file or software before program delivery.

Find out what limitations actually exist regarding the company policy on downloads and software trials. Sometimes the rumors of limits are worse than the truth. Many ninjas run specialty software from their home computers and integrate the outputs into their work programs. I am not advocating that you do your day job on your personal time, but maybe you can work something out with your manager and work from home when you need to use software that cannot be used on your work computer. If you find you are fully locked down to PowerPoint, it could be worse. Check out chapter 10 on all the cool things you can do in PowerPoint that many EHS professionals don't know about.

Working with Your Learning and Development Department

Some trainers reading this book will be working for an organization that is large enough and smart enough to have a learning and development (L&D) department. If you do, please remember its employees are not the enemy. A recent learner at my daylong Safety Training Ninja program shared that she was admonished by her EHS boss when she went to the L&D department for help with a training project. This was a *huge* global company that was attempting to hold all EHS training design, development, and delivery within the EHS department. This ninja thinks that's just making extra work for the EHS staff. Why would we not seek help from education professionals who understand a great deal about learning?

When I am doing professional development at trainers' conferences and share that I am an EHS professional, I always get the same look—that "what are you doing here?" look. Then the trainers inevitably ask: "Why won't the EHS department do what we suggest? Why do they fight the ADDIE cycle? Why do they veer off and do their own stuff outside of the L&D department?" And as much fun as those questions are, when I meet with EHS professionals who have L&D departments, they always ask why the L&D folks are so rigid and won't take the time to understand that there is a particular way we do things in EHS.

It sounds like we have some misunderstandings about each other. I often offer to act as a translator between the two groups, or at least I recommend that the EHS staff attend my Safety Training Ninja program. The above are only some of the problems with these relationships. But there are ways we can improve.

- Learn their language—now hold on, I will ask that they learn yours, too. But this book is for you, not them. You are already well versed in many of the ideas they use and apply to their craft (I hope that by now you agree that great training is a craft, not merely part of a job you do). Letting them know that you want to learn more about ADDIE, that you could use some help on activity design, and that maybe they could sit in on your delivery could lead to great things for your department, their department, and your organization. Throw in a mention of Bloom's Taxonomy, the law of recency, or performance-based smile sheets to let them know you have been learning. Collaborate on the work you can do together, *then* you can explain why you think some of the EHS materials you need to teach won't work in an e-learning format or whatever battle you may be waging with the L&D team. Imagine how much more communication both departments can achieve if they understand each other and each other's language.
- Be cognizant of what a learning management system (LMS) can do. If you have one at your organization, ask the L&D team for help understanding it better. It is likely used for e-learning and face-to-face training data. The LMS is a method for keeping track of all the training conducted by the organization. It could be a robust, expensive piece of software, or it could be a great Excel spreadsheet that some really smart person created. Maybe it's simply a list of training info. Regardless of what you have, try to get a better grasp of what it can do for the organization and why it is used. Note that ANSI Z490.1, section 7, recommends keeping good records of the who, what, when, and where of learning. Proper recordkeeping will also help during consensus standard audits, regulatory investigations, and legal cases.

Side Hustlers

Now I know this moniker might sound bad if you haven't heard the term before, but consider the context. One type of side hustler is that guy who picks you up in his personal car that you requested via an app so you didn't have to take a cab. That guy is doing his side hustle—his job on the side—and he is part of the gig economy. The basic premise involves connecting freelancers with users who need specific work done. There are several websites that connect side hustlers with clients. I have used Fiverr, Upwork (formerly E-Lance), and Thumbtack; but there are others.

How can a ninja use a side hustler? You may need one to develop specific graphics, edit complex photos that you cannot fix yourself (such as photoshopping guardrails into an otherwise great shot), or maybe you need to hire a freelance ISD professional to help you design a great performance-based smile sheet. The process is pretty painless. All you have to do is sign up, submit your request for work, and then set the parameters for what type of person you want to do the work. You can search for specific skills or software capabilities, physical location, or level of experience, and you can read reviews from previous clients.

Remember that in a previous chapter I mentioned I used a macro to produce my learner manuals. I used Upwork for that. I gave them an example of what I wanted it to look like, how much I wanted to pay, and what computer system I used to ensure it would work properly. I had done similar projects in the past, and although I got a great price and quality work from freelancers from distant parts of the world, I found trying to stick to my time zone and continent made it easier for me to work during normal business hours. All of this assumes you have a budget. If you can convince your boss that you can produce an amazing learner manual design for under US$200 (minus print costs), could you get that approved?

Others

Here is a list of people, groups, vendors, and others who can help with your training program. Find out how you can put their expertise to work for you.

- LMS vendors
- maintenance companies who service equipment

- PPE providers
- paid training consultants
- e-learning providers
- downstream contractors
- upstream contractors or owners
- trade associations
- professional associations
- internet searches (don't forget my earlier caveats)

Make New Friends

When I was at university, hardly anyone had computers of their own. I was lucky to have access to a laptop, but systems weren't in sync like they are today. Every project I created at home and brought to the school computer lab to print was on a floppy disk—the actual floppy kind! Regardless of how I tried, my projects were never easy to print. Formatting was lost, and headers, footers, and bibliographies just would not print. It was a nightmare until I discovered a secret (at least to me). Make friends with the IT staff. I explained the challenges I faced and asked for advice. The IT guy would review my files and help me out. The second time I needed his help, I planned ahead. I brought cookies! Not homemade, but a package of nicely wrapped cookies that came stacked four high. You may know which kind I mean. They were a bit expensive—just enough to show I cared, but not enough to hurt a university student's budget.

the gift of cookies

I must admit, I learned this ninja trick from my very good friend Pandora (no comments about the name!). She taught me that simple kindness toward someone who is helping you beyond the scope of their job should be thanked at a minimum and "cookied" for even greater help. If you don't bake—or aren't sure of a person's gluten tolerance—gift cards from coffee shops are great thank-you gifts. Think about this simple gesture after the IT team helps you gain access to online storage or a free download. They didn't have to be nice; they chose to be.

References

Escribano, J. C. 2009. "Learning Circuits Big Question: Working with Subject Matter Experts." LifeLongLearningLab. Accessed August 31, 2018. https://mylifeismylab.wordpress.com/2009/09/28/learning-circuits-big-question-working-with-subject-matter-experts/.

Chapter 9

Integrate E-Learning into Your Program: More New Ideas

Why and How to Use E-Learning in EHS

EHS instructors love awareness training, and e-learning can be perfect for that: short bursts of microlearning on required topics that you don't want to develop into a full face-to-face (FTF) class. E-learning is also great for follow-up training, refresher training, or on-the-spot training. There are many ways to deliver e-learning, from complex programs with comprehensive interactivity to simple videos accessed via QR codes available right at the location of the hazard or where the learning needs to take place.

Although most of this book is dedicated to FTF, e-learning can and should be a big part of your training and education solutions. This chapter discusses some of those alternatives but also brings the ADDIE systems approach to designing e-learning. Before we can get into details, we need to understand the two core categories of e-learning (see figure 9.1).

- Synchronous e-learning occurs when learners attend online training at the same time that it is facilitated live by an instructor. Examples include webinars and online training with interaction, including white boards and chat rooms.
- Asynchronous e-learning applies to learning when the learner completes training without a facilitator present. Examples include,

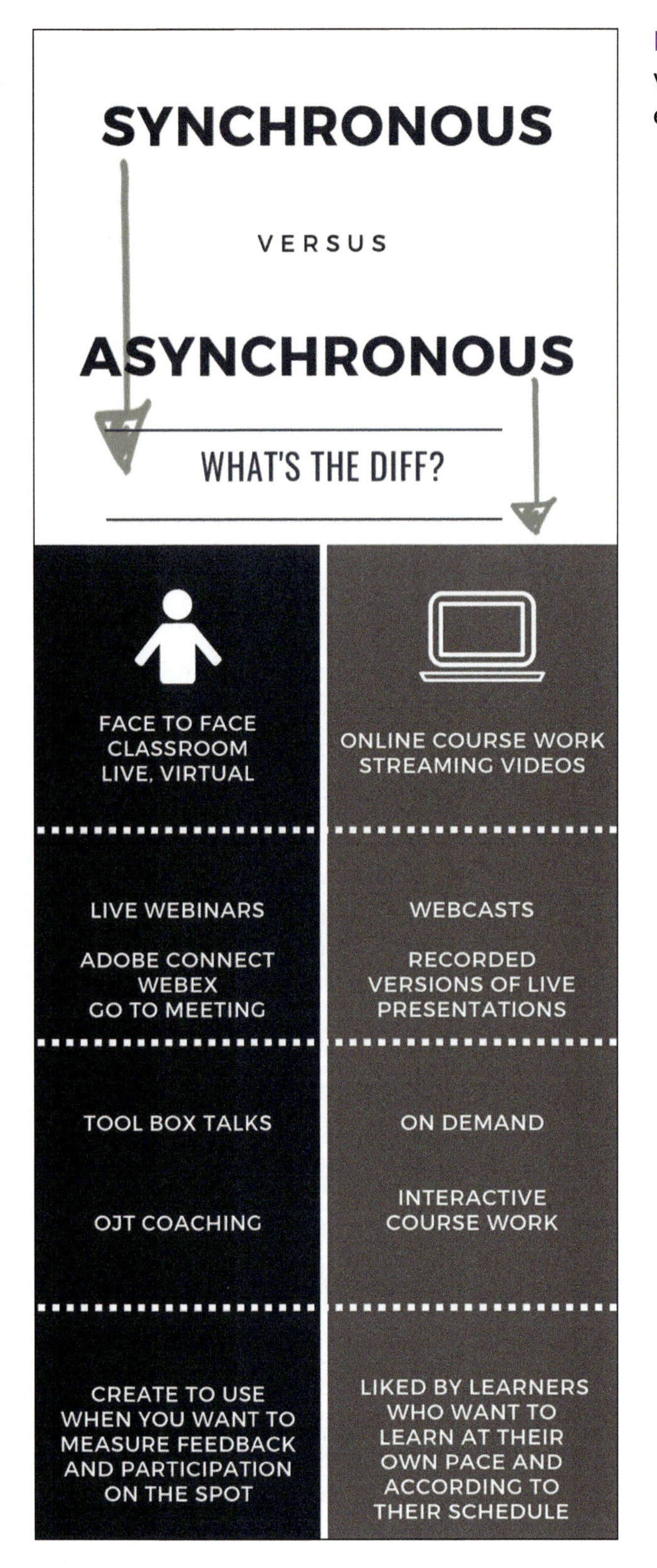

FIGURE 9.1 Synchronous versus asynchronous e-learning

but are not limited to, on-demand videos, online interactive training courses, or previously recorded webinars.

Learning Management Systems

If you have a learning and development department at your organization, you probably already know what a learning management system (LMS) is and probably have had to use it. An LMS can be a simple spreadsheet that tracks all your training delivery, or it can be complex, vendor-provided software that can help you author and distribute your e-learning materials. There is no point in offering online training if you don't know who took it or whether they were successful at bridging the learning gap. The LMS does that and so much more, depending on the vendor and your budget.

I have often assisted clients in choosing an LMS. They range from free to very expensive. The range of services and options offered differ greatly as well. Free LMS offerings are not usually good for an organization that doesn't have the IT support to run the LMS. This is why the other options that provide support cost money. Look at it this way, you can pay an internal staff person to administer the IT functions it takes to run a free LMS, or you can outsource the whole thing. Either way, there are fees involved. If you are tasked with finding an LMS for your company, ask for help and do a lot of research. Bring your IT and HR staff into the discussion. You will need them to help run it, and HR may want to use it too. Here are some things to consider when searching for the right LMS.

- Number of users. Some companies charge based on the number of learners in the database. Some charge based on the amount of content you load into the LMS.
- Types of training. You will most likely want to develop and deliver training with materials that are compliant with the Sharable Content Object Reference Model (SCORM). Although this is the current industry gold standard, not all LMS offerings use compliant materials.
- Uploading materials. Does the LMS offer real-time uploads that update your materials the moment you make a change? One client LMS I worked with required that I send them the files, which they would upload within thirty-six hours. From my experience, I want to be able to edit and immediately upload any changes I deem

necessary. I don't want to wait for someone else's schedule. Inevitably, I find that as soon as I get the final approved copy of the e-learning program uploaded, someone finds something they want to change.

- Frequency and length of courses. Going back to the first consideration, does how often a course is accessed or how long it is affect the pricing of the LMS? Is there a limit to the size of the course or how often it can be used in a set period of time?
- User savviness and your allowance of "bring your own device" (BYOD). Will your LMS support assisting your learners who are using different devices, or will they limit access to corporate computers? Do they have free tutorials that can be shared with your learners, teaching them how to navigate the LMS initially? Otherwise, you may be developing e-leaning on how to access your LMS!
- How much data you want. Or do you even know what you want yet? If it's the same cost, I vote for access to all the data imaginable. You often don't know what you want until you start digging into the data. Imagine that you are able to analyze the data and determine that 80 percent of learners miss the same questions in your learning program. That's a valuable thing to know—and it probably means you need to look at what you are teaching them. Be sure they offer support to you during the initial data analysis. There's no point in having data that you can't make sense of.
- Customer service. This is a key issue. Can you call for help or will you have to send an email or read webpages filled with FAQs? This ninja needs both. I like to be able to search for the easy stuff, but when I am really stumped, I want a person on the phone to help me—and quickly. Also try to determine if they are used to dealing with non-L&D professionals. If they are speaking to you in terms you don't understand, are they ready to support your learning curve?
- Free trials and sandboxes. Every LMS vendor should offer you a chance to play around in the LMS. This is sometimes called a sandbox. They give you temporary access to upload your programs, assign them to learners, and then have the learners provide feedback on how the system felt to them. I usually ask the most and least computer-savvy people to help me with this evaluation. Often the

most confusing part of LMS management for EHS professionals is the uploading and coding of staff in the appropriate learning groups. I have been completely baffled by some of the products out there.

- Bonus products. Some LMS vendors offer lots of free products along with their subscription services. Some even offer free safety training—but that doesn't mean it's right for your organization. These are called freebies, but if you never use them, do they justify the cost of that LMS over another?
- Their product evolution and what you will want in the future:
  - How often do they upgrade their software and how savvy are they regarding the leading course-authoring software? Some of the great companies have cheat sheets that help ensure you build your course so it seamlessly works with their LMS. Others have told me I need to contact the authoring company for help with their LMS!
  - How long have they been around? Does this company have a good track record and long enough history to ensure they can handle your needs? Cheaper isn't always the best choice.
  - How will this integrate with other software your company uses? Think HR databases of learners' names you will need to enter into the LMS in order to assign courses to learners.

Several LMS providers have great, integrated software as part of their packages, which allows you to develop materials right in the LMS. This can be very helpful in developing your first online learning programs. But please be cautious. If you use their software, which works only with their LMS, will you be able to take the product and all the information with you if you change LMS vendors? In many cases you cannot, and you will need to recreate the training program so it will integrate into the new LMS. Maybe that is not a big deal, or the tradeoff is worth it for the sake of simplicity. The choice is yours; just make an informed one. Choosing the right LMS is neither easy nor inexpensive, and, sadly, I have never met anyone who really thinks their LMS is perfect, whether because of cost or usability. Everyone I talk to wishes it did something more or different—or just costs less considering what they use it for. However, maybe armed with the knowledge above, you will find the perfect LMS for your organization.

Technology and Learning

How you deliver your synchronous or asynchronous programs depends on your analysis. Can you deliver an e-learning program that includes all the ADDIE elements and is valuable to your learners? Whether it is a webinar, a previously recorded webcast, or a comprehensive learner-navigated online program, will the participants walk away from the computer or phone and say, "Wow, that was helpful and not as bad as I thought it would be?" E-learning development and delivery tools have improved at a staggering pace in the last five years. I know this because I can author content I would have had to pay someone to do for me only a few years back. Before you can start implementing ADDIE, you need to know what ninja tools are available and how you can use them to make your job easier and the training better.

Content-Authoring and Video Software

Years ago, ISD professionals had to hire programmers to take their e-learning from creative concept to a flash file that would play on many devices. But some videos would not play on some devices, and you would have to miss that really great cat video. That's because some things played on Apple products and other things worked only on Android or Microsoft products. This was bad for designers, learners, and cat video lovers. I could explain this in much greater detail, but I am sticking to the rule of telling you only what you need to know. Suffice it to say that a solution was created that all platforms would accept: HTML5. Now it doesn't matter what software you use to create a product or what device you use to watch it. If the product is produced in HTML5, it should work seamlessly across all devices.

Why do you need to know this? Because if you are developing e-learning materials, you must ensure your output is HTML5 if you want to record the data in your LMS via a SCORM-compliant system that works on all devices. Being SCORM compliant means that the products you develop in one SCORM-compliant system will run in any LMS that is SCORM compliant.

What is authoring software and how can a good ninja use it to make some great products? There are several products available for the different types of e-learning you are developing. I will touch on a few of them, but use your ninja skills to research others. Table 9.1 lists some resources I've discovered. Most EHS pros I know use the options that follow. It is reasonable that by the time you read this book, even more cool stuff will be available.

These are complex programs that may challenge seasoned EHS pros. If you work with one often, you can excel at using it. I will be honest: I usually hire college students to help with my e-learning projects, and they master the software quickly. However, I needed to take a three-day course to feel confident using Captivate. I used this exclusively for several projects and honed my authoring skills on this program. And those same skills transferred nicely to Articulate, which feels pretty similar to use.

One ninja trick I learned quickly is that you need two screens to run the more robust software. There are so many options and tools to use while developing your product that you can't see them all on one screen. I invested in two large monitors and don't even try to work from my laptop screen. I believe it will continue to get easier and easier for non-ISD professionals to design and deliver courses using these products. The good news is that once you learn one, the others make sense since some of the design concepts are similar in any product. Think of it like photo-editing software for your phone: The various apps can be really different, but you can usually figure them out since the basics are all the same.

Articulate

This is one of the top HTML5 authoring tools. This robust software allows you to create rich, interactive e-learning. Articulate 360 is a suite of products that work together, letting you make and distribute some of the coolest and advanced training programs you can imagine. You can branch learning concepts, allowing for all kinds of learner-controlled navigation of materials to meet each person's own pace and needs. But to achieve all that glory requires you become a ninja with that software. It is also a space hog on your computer and is constantly updated to keep the glitches to a minimum. I used this software for a big project that was mostly a templated design with data entry and use of the quizzing options. I like the look and feel but have yet to truly design in it. I have trusted ISD colleagues who swear by its superiority. The downside is their customer service; it is almost exclusively via email, so if you have a technical problem, it will be a while before you get it fixed.

TABLE 9.1 Cool Resources: Free or Low Cost

Product Name	Cost	Download Required	The Cool Thing It Does
Articluate.com (the website, not the software)	Free	No	Even if you don't use the product, these folks share so much helpful information that you should use it as a resource.
Audacity	Free	Yes	Best and simplest audio editing; great for beginners and pros alike.
CamStudio	Free	Yes	Records screen activity and allows editing. Simple program and intuitive.
Canva	Free unless you buy add-ons	No	Take all their sample templates to create simple graphics that will make your information look better. I use it for social media as well!
Doodle	Free unless you upgrade	No	Nice to use for surveys or questions to learners or SMEs. Similar to other survey tools.
GoAnimate	Free trial for fourteen days	No	A cloud-based, animated video-creation software that is pretty easy to learn even if you don't have background in animation.
Polldaddy	Free with limited use	No	Great way to send advance polls, or use it to conduct evaluations. Be sure you can integrate results into your LMS if that is your goal.
Powtoon	Free with limited options	No	Another cloud-based animation program that is easy to learn.
Screenr	Free	No	Web-based screen recording for demonstrating how to navigate new software or a website. Limited to a five-minute screen capture (pure genius!) and no editing allowed. Get it right, or you better enjoy your errors and imperfections.
Skype	Free	Yes	Screen sharing for webinars and free or darn cheap phone worldwide.
Survey Monkey	Free with limited options	No	Similar to Polldaddy.
Wideo	Free for seven days	No	Cloud-based product to make short "explainer" videos with templates to help you create them.
YouTube	Free	No	Surely you know how cool YouTube is.

Captivate

Captivate is part of the Adobe software family and also offers rich interactivity. Captivate has most of the bells and whistles that Articulate provides. Captivate offers an online subscription service so you always have the latest software. Because they are part of the larger Adobe company, you can call customer service with questions, which is nice if you are in the middle of a project and want to keep developing. Which software you choose often comes down to what you are familiar with or if you have a mentor who prefers one over the other. Many ISD professionals have worked with both and usually prefer one, but they are happy to work in either.

Camtasia

Camtasia is video-editing software and is used by many ISD professionals and several EHS pros as well. Yes, it enables you to edit video, but what I love is its simplicity; I don't find it as overwhelming as some of the other video-editing software. I like how easy it is to drop in film clips, cut and trim them, and then add any audio or background music.

Another bonus is that it can record what is happening on your computer screen. Imagine you need to do a quick, five-minute video on how to enter information into the new near miss reporting software your company just purchased. Instead of dozens of boring screenshots, you can actually demonstrate how to do it while operating the software. You can edit out errors and slow connections and end with a small, chunked video that tells learners only what they need to know. Camtasia has a little sister called Snagit, which can do much of what Camtasia does on a smaller scale and may be just what your organization needs. Snagit was originally developed as a screenshot program but has morphed into much more. Snagit is still my go-to for screenshots and simple editing of those images.

Audacity

This is the quintessential audio-editing software. No matter which ISD pros I speak to, they all use Audacity. First, it is very easy to learn. I have made five-minute tutorials (using Camtasia) to teach clients how to edit audio on Audacity, and that's all the training they needed. Second, it's free! Yes, you need to convince IT that being free doesn't mean it carries a virus, but be sure you download from a reputable site. I don't like to create my audio using the options in Captivate, Camtasia, or Articulate because then the audio is

powerpoint shortcut

If I have to put words on the screen during a video, I usually create the slides in my template in PowerPoint, then convert each slide to a jpeg, and drop it into the timeline of any of the above software. I like using the PowerPoint templates because they are already created with the right colors and fonts and are easy to design and develop quickly.

embedded in that software file. When I use Audacity, if I ever want the files for another purpose, they are all saved as MP4 files in a separate audio folder for each project.

Animation Software

If you have the time to learn this software, it can be valuable and add a bit of whimsy to usually serious topics. I have one client who does a new animation every month, and his learners love it. It really personalizes the materials since the main character is styled to look like one of the safety staff members who narrates the program.

Web-Conferencing Software

Whether you plan to host a webinar live for your learners or record it and allow on-demand access later, you will still want to be sure you are using the right software to achieve your learning goals and bridge the learning gap. There are several different web-conferencing software products that offer presentations, group chats, participant white boards, breakout rooms and more. The more features you choose to use during your programs, the more complicated it will be for you and your learners. This ninja likes to evaluate the three stages of a webinar to determine which is the best software for her clients to consider:

- Before webinar
    - Maximum attendance limits. Some software cannot support hundreds of synchronous learners. Too many users can cause slow transmission or a system crash.
    - Global attendance. Do you have global learners, and will the system you use work for them?
    - Content shared during webinar. Will you show only PowerPoint or other simple software, or do you plan to show videos or instruct on how to use software that is running during the webinar?
    - Registration tracking, follow-up, and analysis. Do you need to have everyone sign up in advance, track details about them, send follow-up e-mails, and ensure they attended?
    - Branding standards. Does your organization want the webinar to have the corporate brand as part of the experience?

- During webinar
    - o Presenter tools. Do you need to annotate on the screen, answer questions from chat boxes, and offer polls and quizzes? Can you do this easily without use of a "producer"?
    - o Video recording. Do you need to record your webinar for on-demand access later? What format will it be in after the webinar? Can you edit it if necessary? I used to record the webinars and then export them in a file that I could edit using Camtasia. These files are normally very large and can tax Camtasia a bit, so be prepared to save your work frequently.
    - o Screen sharing. I don't mean just presentation sharing, but showing software in use or playing short videos. Practice going back and forth between what the learners see, and be sure the sharing is as smooth as possible.
    - o Moderator options such as polls, quizzes, and surveys. These tools can really enhance a webinar, but you want to be sure the tools are intuitive for you and your learners.
    - o Webinar features. Chat boxes, white boards for collaboration, and breakout rooms are all cool features, but do you need them? And can you manage the technical components of using them?
    - o Production assistance options from vendor. Will you be doing some really big webinars with some complicated features? If yes, you may need to hire a producer from the vendor. The producer makes sure all the IT runs smoothly so you can focus on content. You can also use another team member to act as your producer, but that person needs to be as good or better at managing the technical elements of the production as you are.
- After webinar
    - o Analytics. Do you need to know who signed up for the webinar, who actually came, if they paid attention, and if they stayed for the entire program? Not all web-conferencing software offers that information, so you need to find out if that is important to your learning strategy and business goals.
    - o LMS integration. Some of the conference software keeps track of all that important data listed above and feeds it directly to your LMS. If that is information you need, this is a valuable service

that only a few providers offer. This may be one of the first features you ask about when selecting software.

- o Editing and posting. If you want to share the live (or prerecorded) webinar with your learners later, be sure you record the webinar through the software, and then be sure the output file works in your LMS. You may still be able to use Camtasia, if you own it, to convert and edit the file, but you want to be sure you are not wasting time and money converting products solely to get them to work with your LMS.
- o Follow-up communications. Will your software allow you to send follow-up communications easily? You will want to get evaluations into the learners' inboxes seconds after they finish the program, whether live or on demand.

Your web-conferencing software may have already been chosen for you by your IT department, or you may need to do research to find the right one for your organization's needs. WebEx, GoToMeeting, ON24, Zoom, and UberConference are all popular platforms, and you need to figure out which is right for your learning strategy, budget, LMS, and the human resources available to produce valuable learning.

There are definite benefits to e-learning and some previously unrecognized disadvantages. Table 9.2 summarizes some of the pros and cons. I often get asked what is the best learning choice for EHS training? The best answer is: sample from all the offerings. No single type of training will work for every organization in every situation. Can you train FTF or do you have a remote workforce? Can you develop the materials yourself, or will you need an expert to tackle the complications of a chemical exposure? Is it best to hire a trainer to get your team to meet minimum government requirements, or will an online program chunked into smaller pieces be better? We have talked about the financial resources and time training delivery uses up. Now is when you determine if e-learning is an option for your team and if you can deliver it effectively to your learners.

E-learning You Can Buy

What about all those companies that offer online training through their sites or can integrate their training into your LMS? Sounds like just what you need? Don't forget about the challenge of training all the learners with all their different

TABLE 9.2 Pros and Cons of E-learning

Pros	Cons
It can be inexpensive compared to flying in remote workers.	It can be very expensive to develop interactive e-learning that may only be used by a few.
Tech-savvy workers prefer this kind of learning to FTF.	Technology may be the enemy, depending on your learners and the technology they have available to them.
You can chunk complex information into smaller pieces that focus on the core information.	It can be boring if you only use the PowerPoint-with-audio approach.
By using an LMS and the right software, you can track testing electronically.	Some learners will do anything to speed through the content, perhaps only skimming content without learning anything.
Great for task-specific training; learners can practice over and over again until perfect.	It may not allow learners to demonstrate complex or life-critical behaviors like they could in the classroom or on the job.
Can be used to share information before an FTF course, saving time in the classroom.	It doesn't allow peer mentoring as easily as FTF course materials do.
Can be used as the test for FTF, OJT, or classroom training.	The authoring software can be very difficult for the EHS professional to learn in the limited spare time available.
Allows learners to move at their own pace.	An unsuccessful first attempt can sour your learners for future learning.
Webinars and webcasts can affordably share specific information among a lot of learners for a great value.	Webinars to large numbers of learners may require a producer to help manage all the details of both the learning and tech sides of the training.

learning styles and experiences with every tool you have. Companies that offer compliance-based (and beyond) training can be a great addition to the company-specific training you may need to deliver. Consider using the vendor-provided program as well as developing the extra information you need to make sure your learners have all they need to bridge the learning gap. I don't think any vendor out there can be everything you need, but you should consider using them for awareness or basic training that you can improve upon.

ADDIE

We have spent a great deal of time and text on ADDIE so far, and almost all the concepts of ADDIE fit nicely into developing or purchasing e-learning. My favorite book for an introduction to e-learning is *Designing E-learning* by Saul Carliner (2006). This is my go-to book on e-learning for new ninjas because it was written before the huge technology burst and the advent of popular course-authoring tools that help EHS pros create e-learning, but not necessarily great e-learning. Carliner's book grounds us in the basics of ADDIE while adding an e-learning twist to program creation. Let's look closely at ADDIE again to show how the elements of the cycle link specifically to e-learning.

Analyze

Before any talk on e-learning can start, ask "*Can* you deliver the program effectively to your learners?" In other words, do you and your learners have the skills and capabilities to use the technology to actually develop, deliver, and learn from the materials?

Learner considerations are:

- Can your learners easily navigate the software to complete the learning? Will you need to train them on how to use software such as an LMS, WebEx, or GoToMeeting?
- Will your learners refuse everything but FTF learning?
- Will your stakeholders see the benefit of non-FTF training? Some industries are much slower to adopt change than others—you know who you are!
- Can you design a program that your learners will want to complete and learn from?
- Can your program meet the needs of most of your learners, making it cost effective?
- Do any of your learners have ADA considerations? That must be designed into the program in some cases.

Technology considerations are:

- Do you have the time, budget, and IT support to generate, deliver, and manage an e-learning program?
- Can the computer on which you are designing the program handle the heavy software demands? I deliver training using Microsoft Surface, but I cannot use any of my authoring software on it—that software is much too hefty. I do all my design work on a system that's as strong as those hardcore gamers use.
- Do all learners have access to computers, tablets, or smartphones to complete the learning?
- Do all learners have access to enough internet speed to complete the course successfully?

- Do you and the learners have the needed hardware and software to successfully complete the learning?
- Do you have or need an LMS to track you learners' participation data?
- Will your IT department allow all learners to have the needed software on the company-purchased equipment?
- Will you have a bring-your-own-device (BYOD) policy? Keep in mind that if you have trade secrets or US ITAR compliance requirements, your learners may not be allowed to use those devices on company property, and you may have to pay them for after-hours training.

Further analysis includes determining how or whether you can adapt any existing information into the new e-learning program. As discussed in the previous chapters, your audience, the work they do, and the learning gap you are trying to fill must be determined before you decide what delivery mechanism you are going to use. This development phase is also when you consider the business goals of e-learning and how you can present any cost savings that might influence the C-suite.

Design

A wise ninja would start small for the first e-learning project. Use the skills you learned in the previous chapter about your design system and then add some extra elements unique to the e-learning, including:

- A more detailed road map. Design your e-learning outline with more detail, despite the fact that it may be shorter than many FTF programs you may have done. The technological elements of the internet, remote workforces, and variable viewing devices can all ruin the best laid plans. Even if your first e-learning course is a simple thirty-minute webinar, you still need to design and plan differently, especially if your audience has never learned this way before.
- A more detailed style guide. Often your marketing and communications departments dictate the PowerPoint templates, but do they have a template for webinars delivered on multiple devices, and do they have templates for use in asynchronous learning developed using Captivate or Articulate? You may need to design one yourself and get it approved. To do that you will need to know the corporate

TABLE 9.3 Comparison of FTF and E-Learning Time (hours to develop vs. hours of delivery time)

	Rapid Development, Simple Projects	Average Development, Typical Project	Advanced, Complex Development, More Media
Instructor-Led Training	22:1	43:1	82:1
Level 1 e-Learning (Basic): Content pages and assessment	49:1	79:1	125:1
Level 2 e-Learning (interactive): Level 1 plus 25% interactive exercises	127:1	184:1	267:1
Level 3 e-Learning (Advanced): Simulations, games, award-winning types of programs	217:1	490:1	716:1

guidelines and the latest in adult learning principles so you can design your program to engage and inspire.

- Stronger design skills. If you are creating course materials in Captivate or Articulate, the template design and "look and feel" can be very difficult to change late in the development stage. In some cases, it can necessitate a total rework. This is not only frustrating for you, but it can also sabotage any money-saving initiative you had pitched to the executives.
- Time. Remember when we discussed the cost of learning creation back in chapter 3? Well, check out table 9.3—e-learning design is even more time consuming. Don't be discouraged at the numbers if you are looking at training a lot of learners. Also, keep in mind that most e-learning isn't delivered in hour blocks. Effective e-learning development can be expensive depending on the tools you use, who you need to train, and how big of a knowledge gap you are trying to fill. A lot of these details should be discovered in the analysis phase, but if you are new to e-learning, remember that sometimes we learn through experimentation and even failure.

Develop

Whatever you choose to use, planning is everything. After you have analyzed your audience and developed your learning objectives, it's time to develop the content. You will need to be even more thoughtful about what is on the screen and what you will narrate. I tend to follow the goldfish rule of memory when designing e-learning: a goldfish has a pretty short attention span—only a few seconds—and you should assume the same is true of your learners. Content should be impactful, images on the screen should change regularly, and interactivity will help learners. Whether synchronous or on demand, I make it a point to redirect the learners' attention every five to seven seconds. You can do that with relevant graphics, changes in speech tone, slide advancement, or more formal interactivity such as a poll, questions, or a quiz on screen.

If you polish your authoring software skills, you can make templates for page types, knowledge-branching scenarios, advanced interactions, and trackable quiz reports you can analyze. Head to the software company's website, YouTube, or Lynda.com for free or inexpensive learning tutorials. I mentioned earlier that I took a three-day class in Captivate to get comfortable with the software. My client wanted a heavily interactive program, and I needed to do more than the typical first-time user would do.

Authoring software is powerful and can overwhelm you as the developer and bombard the learners with far more flash and bang than they need. Worse yet, you might fall into the trap of presenting created versions that are trite and dreadful to sit through. Be careful not to blindly adopt the precreated templates that the software provides; they may not match your brand or the style of learning of your audience. I am definitely not a big fan of the "talking head" templates, which are basically long monologues with the graphics changing to match two people talking about a subject. That's not much better than PowerPoint bullet points with audio added.

The course-authoring software is usually obtained by subscription and updated often to fix glitches, with major program upgrades supplied regularly. If you are in the middle of a product development, *do not* upgrade until you have completed your project, published it, and confirmed it is delivered as you desired. I have had several L&D professionals warn me about data loss, template changes, and total loss of work due to upgrades. I can't swear this would happen to you, but why chance it? The files you are creating within any of these programs are large. Save often and back up to an external drive or the cloud. You will want to be working on the files on your actual desktop and not on those in the cloud. The software gets clunky and closes unexpectedly if you work exclusively with files in the cloud.

Implement

Just as you test run an FTF program, you need to test all e-learning first. If you are running a synchronous webinar, you need to be sure that the delivery is perfect, it can be seen and heard clearly by everyone, and it achieves your learning objectives. If you don't do that the first time out, you will struggle with getting learners to log in for a second try. With new clients, I often conduct a dry run of webinars with a few select learners. They can help improve content, and they can give feedback on the smoothness of transitions, poll questions, and chat options. Be sure your audience knows how to work the tools inside the webinar software since all the software is just different enough to confuse learners not experienced in webinars. If you are delivering the webinar, it is critical you are comfortable with the final version. I have worked with almost every webinar software product out there, and I still always do a brief dry run. With system changes, software updates, and my computer peripherals, there is always something I need to fix in the delivery to ensure success.

If you're working with course-authoring software, there is more room for error and correction before a full program launch. When I use Captivate or Articulate, I have the SME and stakeholders review the product before I place it on their LMS. Once on the LMS, I do more test runs. I even have my staff log in as guests to test it on all types of devices. Does it work on Apple and Microsoft computers? What about Apple and Android handhelds? If your learners will be using their own devices, do you have the staff to help if there are issues with the program running smoothly. If you don't, maybe you will need to insist they only use company-provided devices.

Evaluate

Your system to evaluate e-learning should be based on the same thought process you used for FTF training. What gap were you trying to bridge, and how will the evaluation support proving you achieved that goal? When using quizzes or tests, be sure to require a grade of 100 percent. As mentioned previously, you don't want learners leaving the training without knowing all they need to. Most of the course-authoring software has system preferences that allow learners to either review the relevant slide and then retest, or just retest. You will need to determine what is best for your learners. Tracking

leverage others on your team

Use those twenty-somethings to help you with your e-learning programs. As mentioned before, I use college students for a lot of the computing side of my e-learning development. These important team members have been editing videos on their phones and playing with GarageBand and iMovie for years. The look and feel of these programs are similar to e-learning software, and I am amazed at how quickly they start to use the software. They are also a great resource when designing for the different generations in your audience.

evaluation results can get pretty geeky, and you will need to decide how much you want to track and how you will evaluate the data at the design stage of the ADDIE cycle.

The Challenge of E-Learning Product Management

I have learned the hard way about e-learning program management. There are a lot of files or assets to keep track of, they take up a lot of space, and you will have a lot of versions of the files as you complete your project. Here are a few things to consider:

- Keeping all those files straight. Build your computer file folders at the beginning and develop a habit of arranging all your folders the same way. I have a main folder with the program topic name and subfolders if there are multiple programs within the topic. I try to do this for every project. It also helps anyone else on the team know where files are stored.
- Naming conventions. Get in the habit of calling the files exactly what you plan to publish them as in your LMS. This is actually pretty important since some LMSs grab the file name as part of their listing inside the LMS. Figure 9.2 illustrates the file-naming conventions.

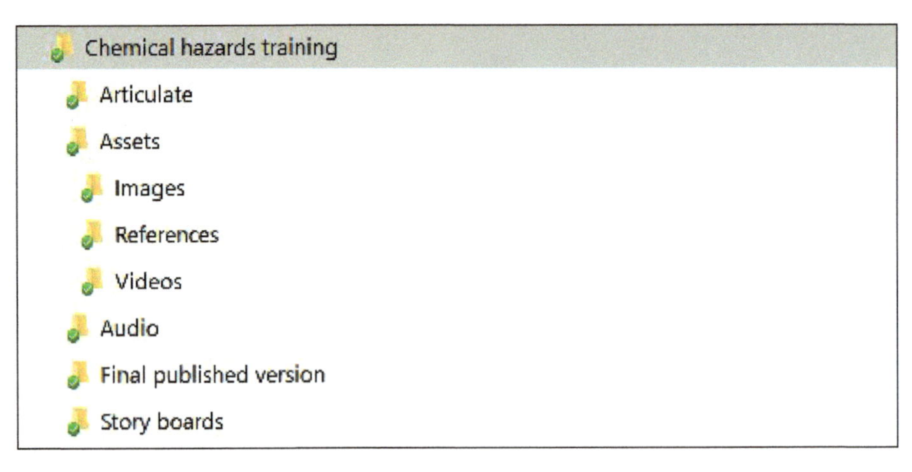

FIGURE 9.2 **File-naming conventions**

- Audio/video. Keep all these files in their own folder so you can find them easily, and give them names that accurately describe them. For audio files, I name them after the slide number they correspond to in the e-learning program. If the audio is broken into several files, I call them something like Slide 2.1, 2.2, and so on.
- Versioning. This can get really tricky quickly. If you and other members of your team are all working on the files, you can get confused about which is the latest version. Use the time stamp on the files to maintain their order and sweep all older files (don't throw them out yet, you never know when you may need them) into an archive. Once your programs are published, you can delete unneeded files.
- Published files. The course-authoring software will publish all the needed files to one folder. There will be a bunch of files that you might not recognize. Do not change anything in that final published folder. Trust that your software knows best. Since you should be working on your desktop during final publishing, be sure you store the final published versions with all your working documents.
- Final storage. Keep all the files in the cloud or on an external drive with a backup as needed. That is a lot of work to lose. Ensure it is accessible to team members if you leave the company. When I do work for clients, I provide my final published files to the client for safe keeping.

Overcoming Opposition to E-Learning

There may be many learners who will resist e-learning completely. It could be they have heard e-learning is boring. Or it could be they are not comfortable with technology and are afraid they will be embarrassed when using the platform. Or they may be accustomed to FTF and just don't want to change. Be kind and help them understand the value and simplicity of what you are offering. Don't get stuck on the generalizations made about certain age groups or industries. People are resilient if you make it easy for them to change.

In the early 2000s, the company I worked for did a lot of FTF training all over the country for our customers. Suddenly budgets were cut, and instructors were told to conduct more webinars. Webinars? For the construction industry? It just won't work, I was told. But we proceeded anyway since we didn't have another choice. We were pleasantly surprised when our first webinar yielded a nice mix of learners. The assumptions about the industry weren't true. Would they have preferred FTF? Probably. But without another choice, the attendance at the webinar was far higher than expected. How did we know they were more resilient than we thought? One of our customers had gathered their construction workers in a trailer and dialed into the webinar from Alaska! They got up early, viewed the webinar as a company, and stayed on the webinar the whole time. And we even got some good evaluations!

The key to introducing e-learning to your learners is to take it slow, make it easy to use initially, and be ready to have private conversations and even some one-on-one training on the technology. Find a technology champion who can help explain how to adopt the new way of learning in a way all your learners will understand.

I could write an entire book on just e-learning for the EHS professional, but this is probably enough for the growing ninja. Refer to the vendor websites, download free trails with permission from your IT team, test your creativity, and explore what feels right for you and your learners.

References

Carliner, Saul. 2006. *Designing E-Learning.* Alexandria, VA: American Society for Training & Development.

Chapter 10
Make PowerPoint Your Best Ninja Tool

In this chapter we will be working on improving our PowerPoint skills. I know there are other presentation programs available, but the majority of EHS professionals use PowerPoint, so I will focus on that. Depending on your experience, you may be new to the latest advances in PowerPoint, or you may have years of experience working with PowerPoint but perhaps are not aware of how advanced it has become. Regardless, this chapter will provide even the most competent user with some great tips and tricks.

If you are ready to jump to even more advanced skills, I recommend going to YouTube and searching for PowerPoint tutorials. If you like a particular user's advice, subscribe to his or her channel so you can see all their videos and get alerts for new programs. I regularly use the skills I have learned on YouTube.

Fundamental Rules of Presentation Design and Development

Think about the following topics before you get too far into the development stage of your program.

Corporate Rules

Consider your corporate rules for PowerPoint and other learner materials. I mentioned this before, but I can't stress enough that you want to be a team player on this topic. You won't be seen as part of the decision-making team if you don't support the corporate brand. Your stakeholder needs to see that you can work within the structure of the organization and that you won't make unnecessary trouble over little things such as fonts and colors.

Number of Slides

More slides don't mean a longer presentation. The length of your presentation should represent the materials needed to achieve your business goals and learning objectives. I have had clients force me to put way too much info on slides or to add more slides because they thought a high slide count translated into a long program. This is not true, but pick your battles with stakeholders. You can convince them later after you have shown them the other great things you are doing.

Fonts

Use a big font. You may need to get the marketing people to sign off on their standard template, so help them understand why you need it. Some good points to make are: "I am teaching the materials in the field with lousy screens and low-lumen LCD projectors, and a bigger font is the only way all the learners can see the materials," or "I want to limit what I put on each slide to the 6x6 rule and I want to emphasize those words."

Color

Use color. But exercise caution, ninjas. I don't mean go crazy. Keep the colors within the corporate guidelines and use them for a purpose, not just because you think they look cool. If there are no corporate guidelines, do an internet search for recommendations of good color schemes, or use the built-in schemes within PowerPoint. Always avoid visually nauseating choices.

Other Basics

- Use key words on the screen.
- Follow the 6x6 rule: Use a maximum of six bullets with no more than six words each.
- Get rid of the sentences. Remove all the little words such as *and* (use the *&*), *the*, and anything that makes a nice sentence but a terrible bullet.
- Try to use symbols whenever possible. They will yield more white space on the screen. Just be sure your audience knows what the symbols mean.

Pictures and Graphics

I talked about copyrights in chapter 5, but I cannot stress them enough. Stealing is wrong! If you are not sure what is usable and what is not, consult with your organization's legal team.

While I'm developing content for my program, I develop a short list of pictures I will use to illustrate certain safe/unsafe behaviors, processes, equipment, and so on. I then ask the stakeholders and SMEs to help me get the images from their sites. If you give someone enough time, it's easier to get their help. The key to using pictures is to ensure they are valuable to the learners.

- Be sure they are clear, in focus, and easy to see from any part of your classroom.
- If you are showing hazards from within your organization, think carefully about how you use them. Can these images be used against you in litigation?
- Fully explain the safe alternative to a hazard. Ask yourself if any of your learners will be embarrassed because you are showing hazards from their work area?

If you have to find pictures and art yourself, be careful of free sites and even what is provided as part of your Microsoft programs. These images are often free for *noncommercial* use. They usually mention the term *education*, but they don't mean workplace education. They mean they're free for public school use. Even if you work for a nonprofit or the government, you probably cannot use a lot of the freebies.

My favorite image site is 123rf.com. They have zillions of images, and most cost $1 to use in a PowerPoint presentation. Be sure to read the usage policy for the image you like to ensure it is OK for you to use. I like images that evoke feelings or address a greater concept. I do searches on risk, safety, environment, hard hats, machinery, trucks, insurance, leadership, team building, and anything else that might work for my program. Will you find great safety training–related photos? Nope. You have to take those yourself, and we know sometimes it's hard to find the "good example." Either fix those images in Photoshop to make them look safe or show them as bad examples.

I once did an hour-long presentation using just pictures and nothing else. It was not on technical training, so it was a bit easier to do. Imagine using only large pictures of each globally harmonized chemical safety pictogram. You don't need words on the screen to explain what each image represents. Instead, ask learners to tell you what it should say to start a discussion on what each pictogram means. Avoid using little pictures no one can see because you think you need to make the slide more interesting. Either make the image large enough to see and teach about, or make your words tell the story. Teach—don't kill them with bullets or lousy images.

Can you take inspiration from one graphic and create a better one for you learners? Sure you can! Canva offers great templates that are free or inexpensive that can make your job aid or process flow appear þetter on screen than something you created with boxes, lines, and a lot of text.

Tables and Charts

Most tables and charts will *not* show up well on slides because they consist of a lot of text and small font sizes. So why not provide these as handouts or job aids instead? Unless there are only a few data points that are clear to everyone in the class, skip the apology, "I know you can't really read this but . . . ," and provide the information another way.

Animations

Unless you are trying to show a particular process, try to avoid PowerPoint-generated animations, except an occasional one for visual interest. Words flying onto the screen and disappearing images can be distracting. It also can take a lot of time to master the flow of an animation, and sometimes you don't have the time to spare.

White Space

Retain as much white space as you can. Cutting back on those extra words and separating one busy slide into two slides will help. The only time you should fill up all your white space is with an image. Sometimes I like to take an image and "wash" it out (put a white box over the image and select transparency); this serves as a subtle background to further support the message.

Transitions

There are several to choose from, but I recommend you stick with the subtle offerings for most of your slides. Save a flashy transition for when you move from one major learning topic to the next or to highlight that a group activity is about to occur. The choices of honeycomb, glitter, and vortex are really cool, but after the fourth slide, your learners will be overloaded with graphics and will lose focus on the content.

Make Learning Sticky

Don't forget your ALP when you are creating your PowerPoint slides. Use your visuals, SmartArt, and pictures to make the concepts you are teaching stick in your learners' brains.

Consider Your Group Size

Think about how many learners will be in the class, where you will be teaching, and if you should be using PowerPoint at all. If it's a group huddle or toolbox talk, maybe a great job aid is a better idea. If it's a huge group, you are going to need a big font so the folks in the back can see your materials.

PowerPoint Skills

Listed below are screenshots and information about the primary skills I believe all EHS professionals should be proficient at to save time and money in your training program development. Please note that I am using Microsoft Office 365 via subscription and that your screens or capabilities may differ based on which software you use.

Edit and Take Advantage of the Quick Access Toolbar

The Quick Access Toolbar (QAT) is a small, almost hidden toolbar that allows ninjas to customize what they can do in PowerPoint with just one click. I created my custom QAT in about fifteen minutes, and it includes all the PowerPoint options I use the most. I also placed some cool buttons there, ones I might otherwise have trouble finding in the myriad of options PowerPoint now offers. The default location for the QAT in Microsoft is above the toolbar at the top of the screen. However, I like to bring my QAT below my ribbon bar; look for that option on the QAT menu dropdown.

Some of the icons I put on the QAT are easy to find in the ribbons and some are not, so for this ninja, creating my own shortcut tool is the best way

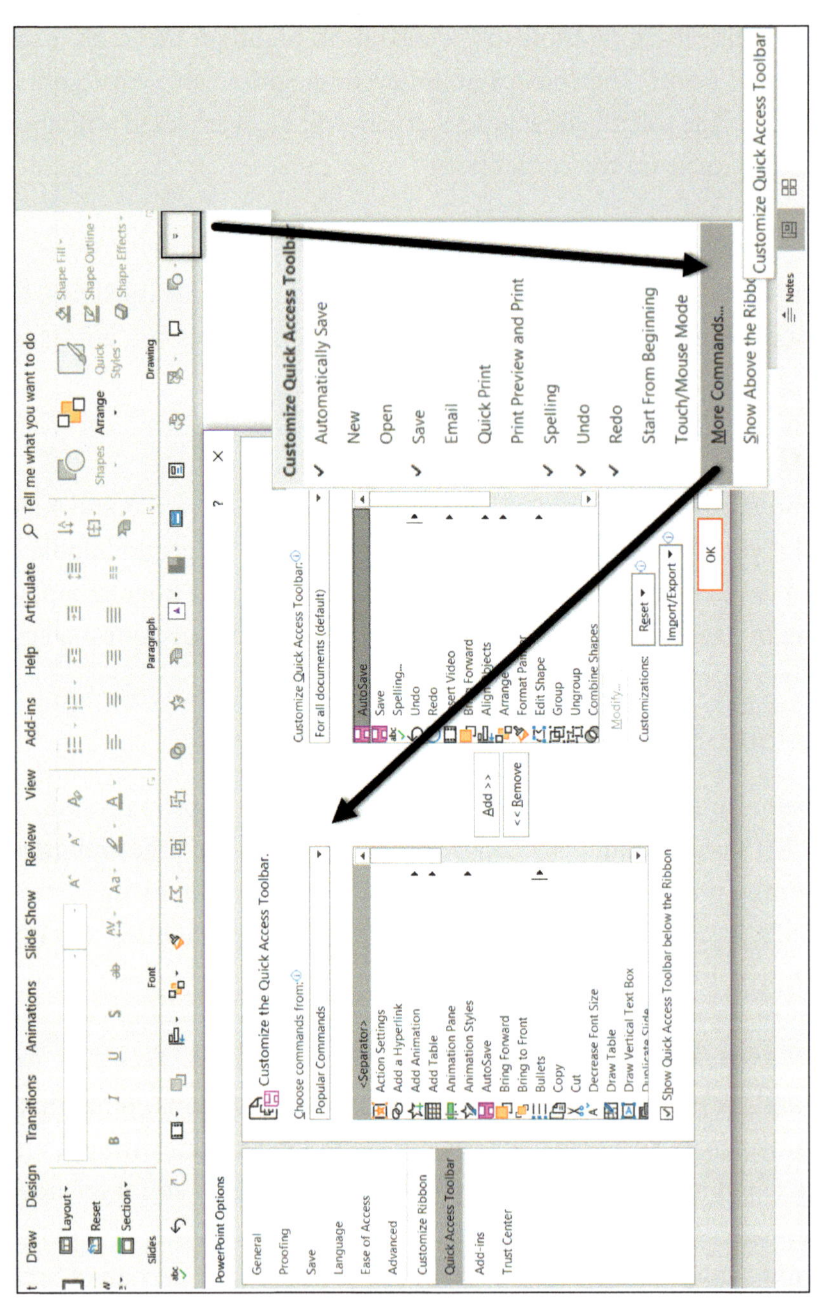

FIGURE 10.1 **Creating your QAT**

to save time and mouse clicks. Figure 10.1 shows how to customize yours. It's a good idea to think of the commands you use most frequently and make a list of those so you know what to add. You also have to know what PowerPoint calls those commands to search for them in the All Commands dropdown menu. Another way to find the command is to choose the Popular Commands dropdown, but this doesn't include everything that PowerPoint can do, only the popular items.

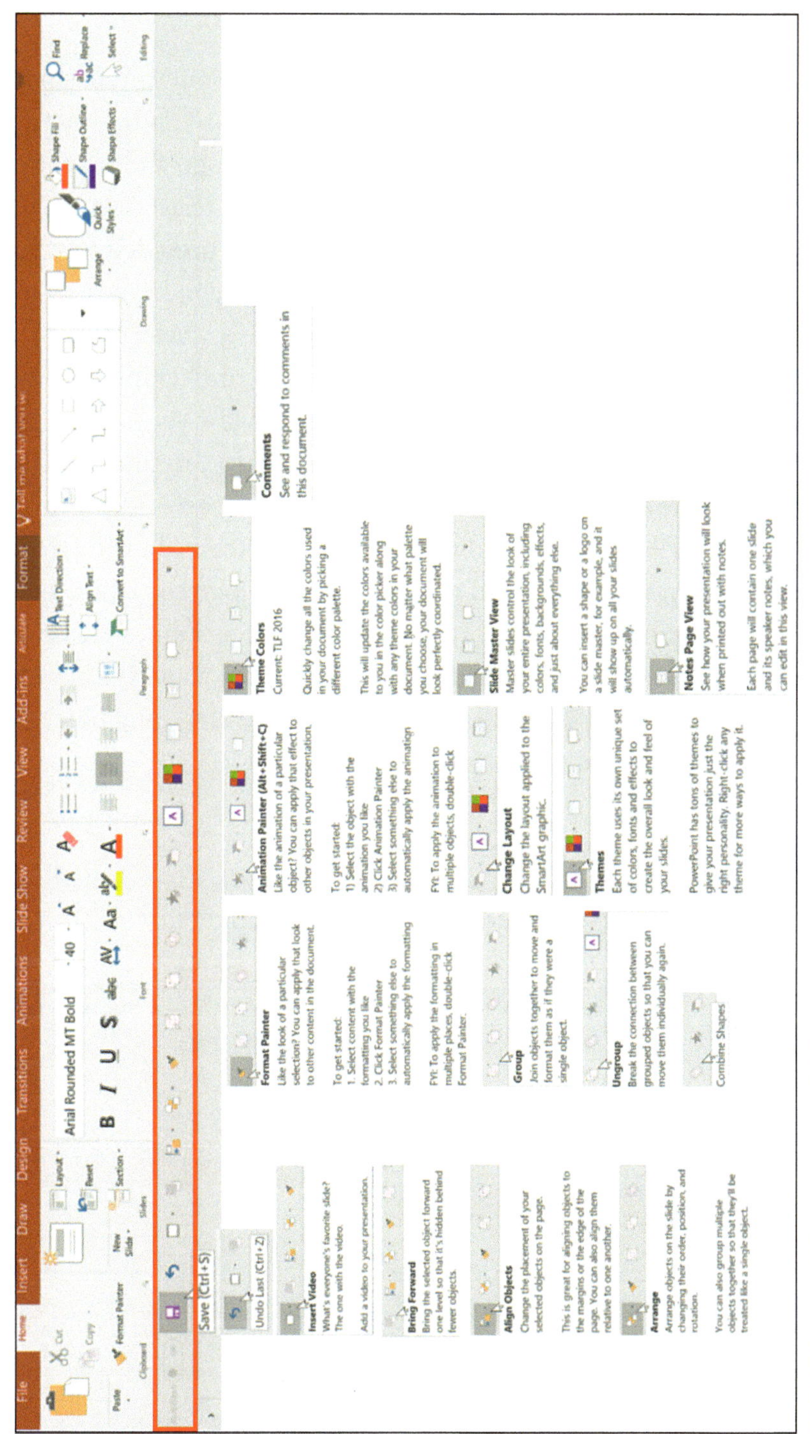

FIGURE 10.2 What's on my QAT

Figure 10.2 shows what my QAT looks like today. I have saved tens of thousands of mouse clicks by using my QAT. You can also create QATs in all the other products in the Microsoft family.

Developing Templates and Themes

If your organization has templates already designed for you, you may not need to read this section. Honestly, most of my clients give me templates that are not really templates—usually one slide with the company logo and maybe some company colors. In this section I will show you how to create, rename, recolor, and design your own template, or enhance the choices provided to you.

First, I click the Slide Master view on my QAT. If you haven't saved that to your QAT, go to the View ribbon and click the Slide Master icon. Within the Slide Master mode, you can create or edit an existing template you chose from your Design ribbon. Inside of the Slide Master view you can change the font and color scheme and add in logos or other creative designs. The value of the template is that once you create all the different base slides you need, you can use PowerPoint to your advantage and grab the right slide from the template. No more cutting and pasting the same logos on every slide or performing other manual tasks. If I am designing for a client, I create a new template with their logo, colors, and anything else they want. If I have five learning objectives, I might create a theme where each learning objective's content gets a slightly different look.

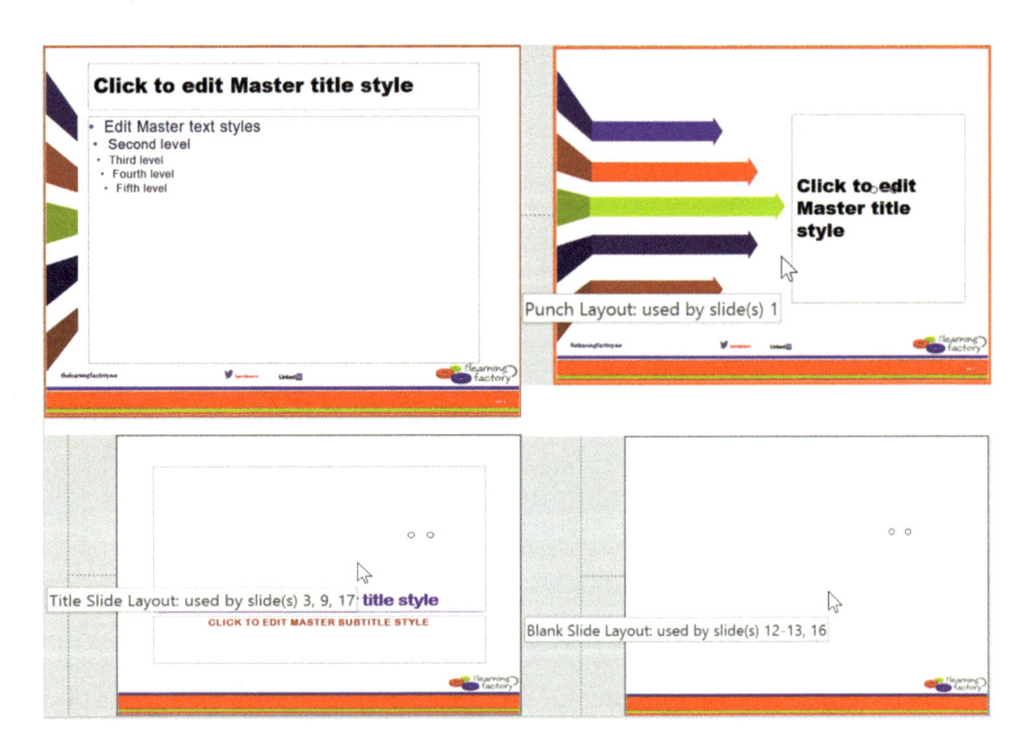

FIGURE 10.3 Sample template slides

Figure 10.3 shows some samples from my corporate template. I have about ten more basic slides that I can choose from, depending on what's going to happen on screen. I name my template slides so they are easy to recognize. Just right-click on the template slide and choose Rename to customize your slide deck.

Once you have crafted your templates, you will rarely need to use new text boxes. You want to avoid adding text boxes in the slide deck. Because they are not part of the template, they will have to be edited and formatted every time you add one. PowerPoint is smart, and it will do a lot of the work for you if you let it.

To save time, look for premade templates in an internet search. Be sure you are downloading from a reputable company, that the templates you select are not copyrighted, and that they are slated for commercial use.

The punch layout slide in figure 10.3 would have taken me hours to visualize and create. But since I bought it from a designer, I just changed the colors and styling a bit and added it to my choices. I recently spent $100 for more than twenty templates. Each template had about fifty slide choices and color variants. The choices are awesome, but I felt overwhelmed and needed a way to easily locate what I was looking for. I grouped each slide by its "content data points," meaning it was a design with five data points versus four or seven. I keep all the 5s together in one folder named 5s. I don't store them by theme but by data points. If I have content that needs to look really good and has five bullet points or concepts on the page, I go to my 5s folder and choose one that fits my program. Figure 10.4 is an example of how slides with five concepts can look entirely different despite being focused on only five data points.

It's not likely that these will work for all of your safety training, but if you need to present data to the C-suite about recent safety successes, wouldn't it be great if your PowerPoint looked impressive?

Changing Themes and Colors

While building your template, you should be sure your colors are correct when you access your template later. Ask your communications or marketing team for the corporate colors. If they provide them as RGB numbers, it will be easy to plug them right into the PowerPoint custom color designer. If they are provided as CMYK or Pantone or Hex, you will need to head to the internet and convert them to RGB numbers. Just enter what they gave you into a search engine. For example, they give you Hex #00FF00. Type in "convert Hex #00FF00 to RGB." You will find websites where you can enter the Hex number to get the RGB number.

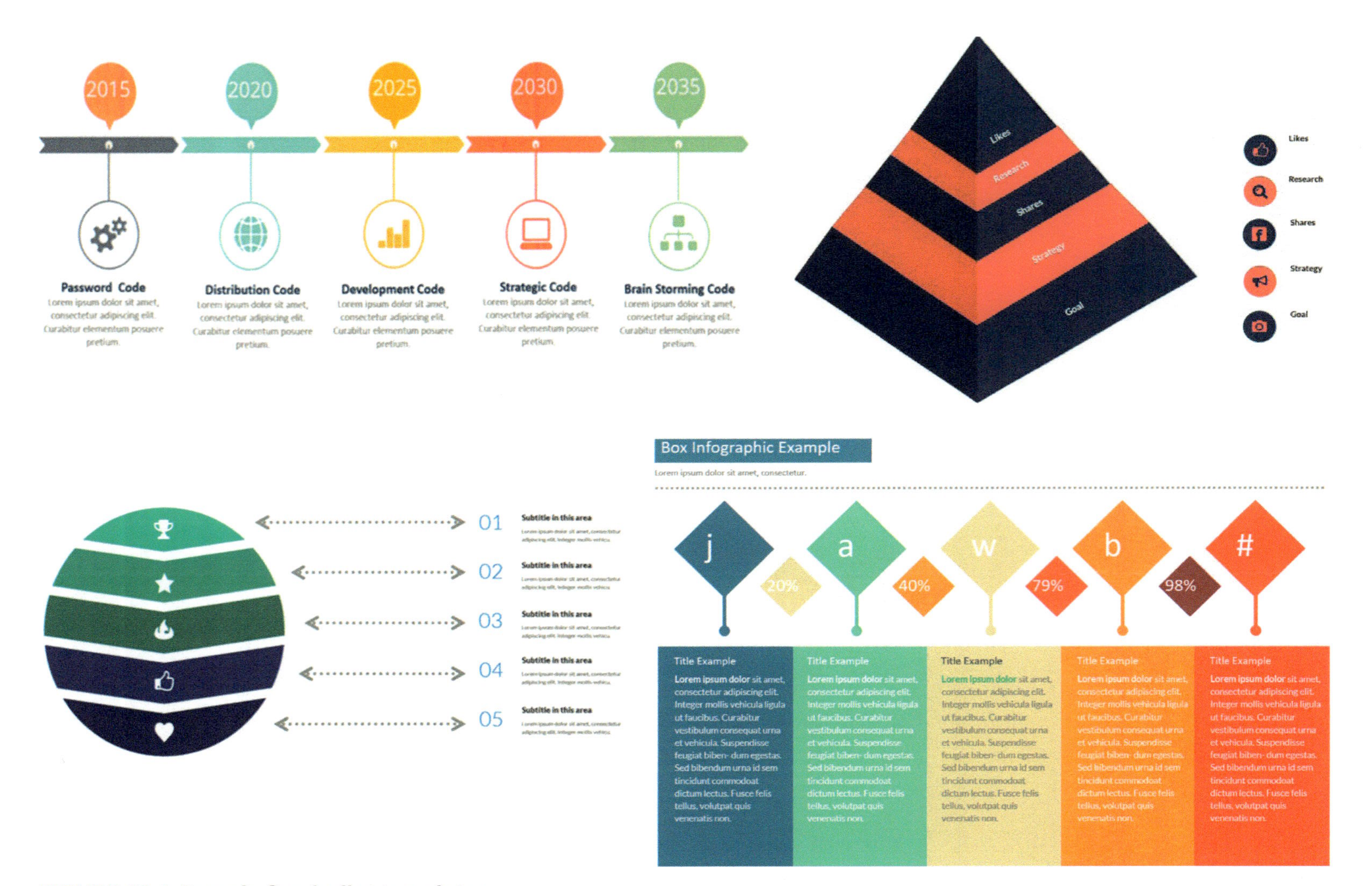

FIGURE 10.4 Sample five-bullet templates

The images in figure 10.5 show how to change the colors of your theme. After you set up the colors you want, be sure to save the scheme with a name you will remember. The next time you need those colors, they will be in your template.

When it's time to finish your template, be sure to save it as a template file (.potx), not a presentation file. This will allow you to select the template you created when you choose themes the next time you start a new presentation.

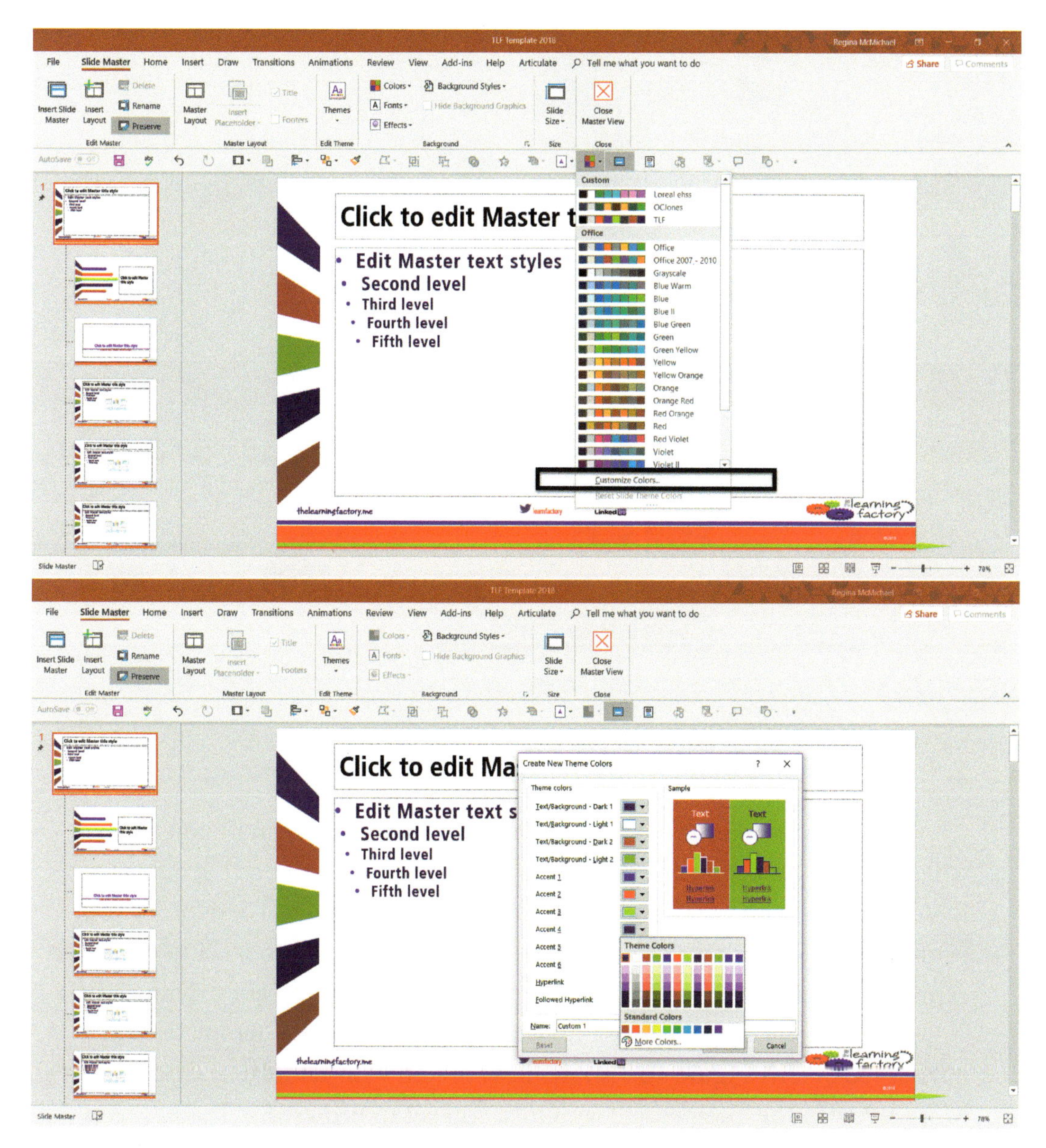

FIGURE 10.5 (a) and (b) Changing colors in your templates

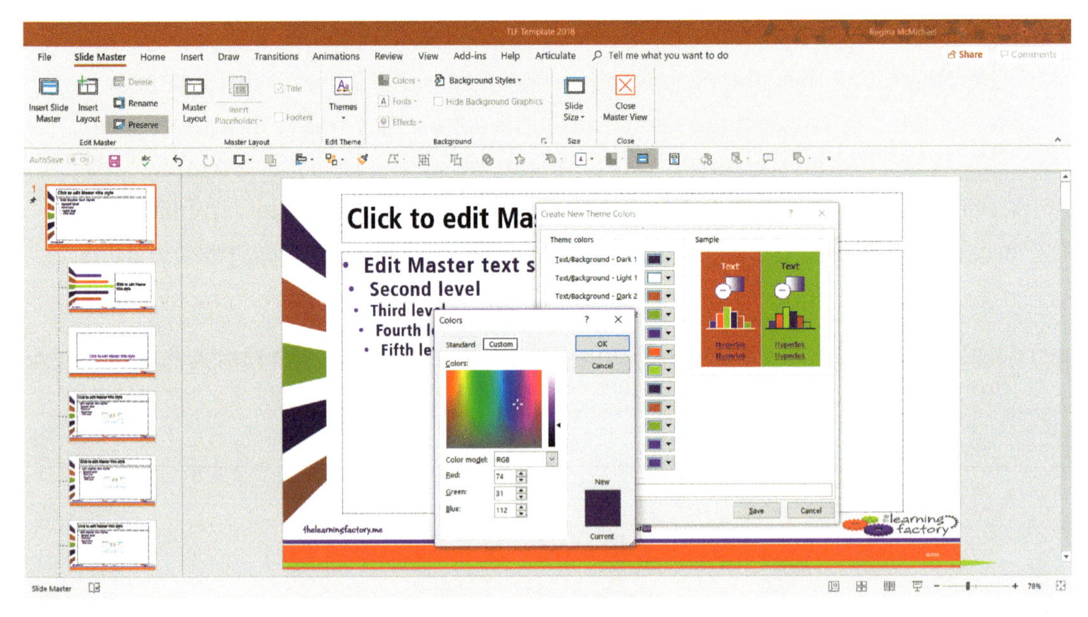

FIGURE 10.5 (c) Changing colors in your templates

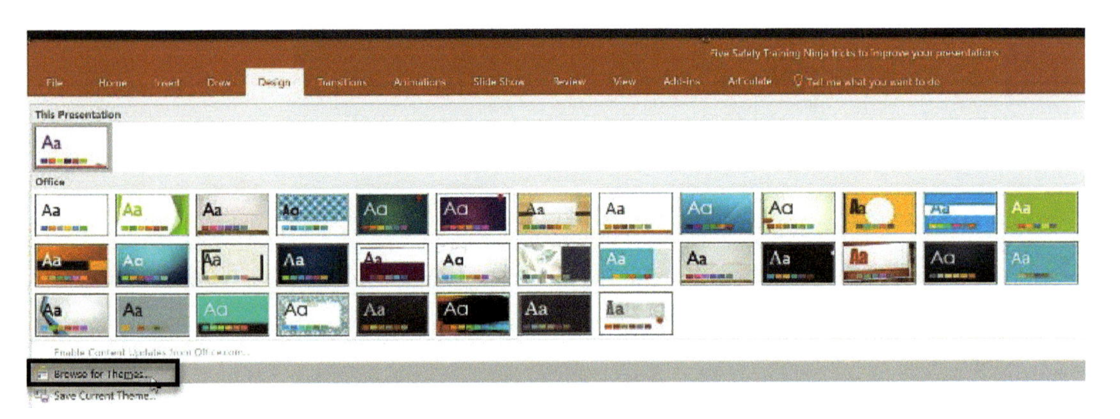

FIGURE 10.6 **The Design ribbon**

Not ready to design your own themes? Click on the Design ribbon and you will see all the choices provided by PowerPoint. Keep in mind that although these came free with PowerPoint, it doesn't mean they are good designs for your learners. PowerPoint offers some visually overwhelming options. Plus, using a canned template doesn't make you stand out as the ninja you are striving to be.

The Design ribbon is the same place you go to find the templates you have saved before. Figure 10.6 shows how to find your templates. Click the More arrow in the Themes gallery. Once you get to Browse for Themes, you can locate it in your saved files.

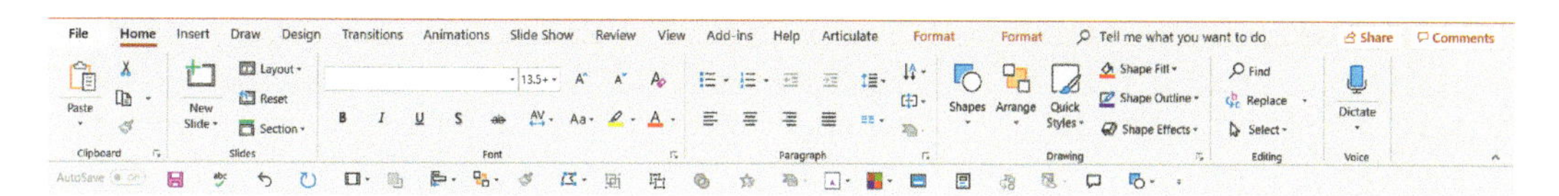

FIGURE 10.7 The Home ribbon

Your Home Ribbon

Right at the top of your PowerPoint screen is your Home ribbon, and it displays a lot of what you'll need to work in PowerPoint (see figure 10.7).

Here is a quick overview of what you can access in the Home ribbon:

- fonts
- font colors
- cut and paste
- bullets
- line spacing
- text alignment
- basic shapes
- shape formatting
- SmartArt (more on this later)

Remember that you can preset a lot of these designs in your own template. Keep reading to find out more on how to do that.

Using SmartArt

One of my favorite tools that is unknown to many EHS professionals is the SmartArt option on your Home ribbon. SmartArt takes simple bullet points and easily turns them into a more visually pleasing and engaging graphic. Figure 10.8 shows how to access the SmartArt options, and figure 10.9 shows the slide before and after using SmartArt. Consider these recommendations when using SmartArt:

- Be thoughtful in your use. A little goes a long way.
- Try to choose a few favorites and use these regularly instead of a different one each time.
- Be sure the graphic lends itself to your bullet points. If there are a lot of words, most of the options don't work.

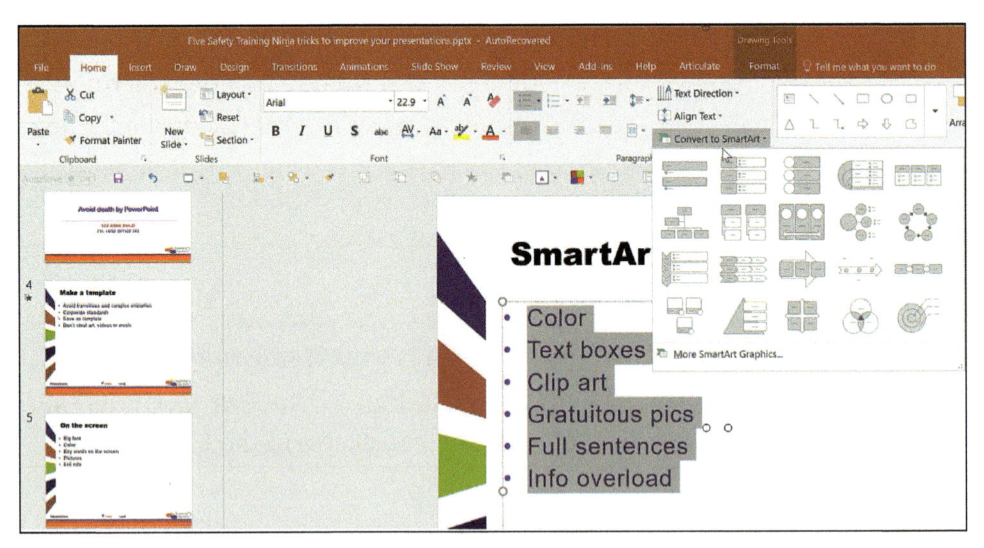

FIGURE 10.8 SmartArt on your Home ribbon

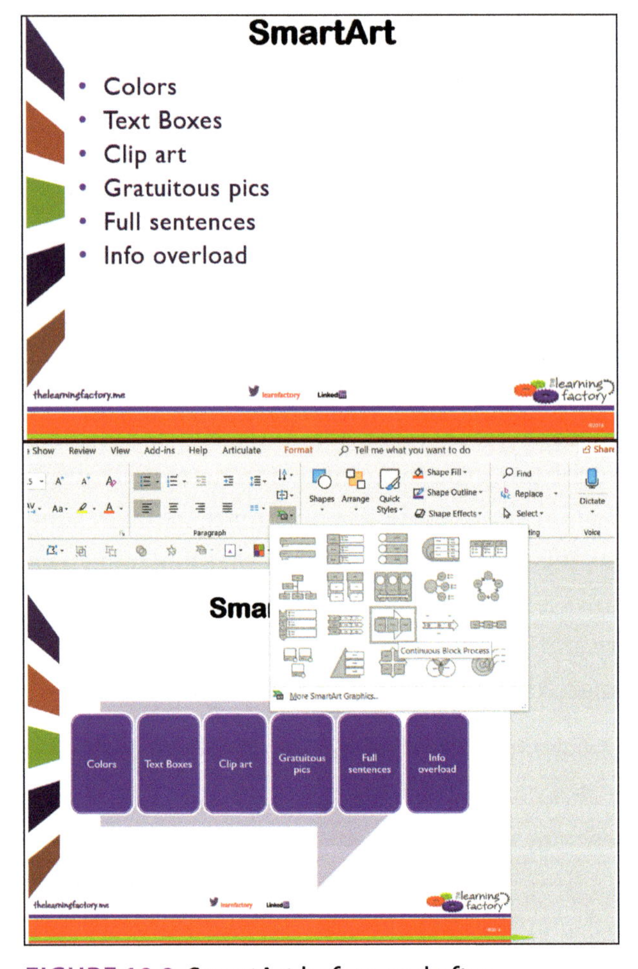

FIGURE 10.9 SmartArt before and after

- When you try to resize or manipulate the SmartArt to fit your needs, it doesn't always work. You will need to practice ungrouping the elements and regrouping them after you have made your edits. This is when my QAT comes in handy, since all the tools I need to edit SmartArt are right there in one click.

Once you have chosen the SmartArt style, you can stay in that ribbon to edit the colors and style schemes. The colors available will match up to the colors you have locked into your template. Although there are several styles to choose from in the options, pick one style and stick with it. Think about the PowerPoint rules: easy to read, visually pleasing, and right for the information presented.

Editing Images

I'm going to reveal a ninja secret here: I use PowerPoint to edit all of my images, for any reason, for any output. It does almost everything I need. Some of the available photo-editing software is just too complicated for this ninja to learn. Editing your images in PowerPoint is illustrated in figure 10.10; it requires that you engage the Format ribbon. This is tricky because that ribbon does not automatically show at the top of PowerPoint unless you double-click the image itself. My favorite tools are laid out below with some tips. If you have never tried to use the image editor, just open PowerPoint,

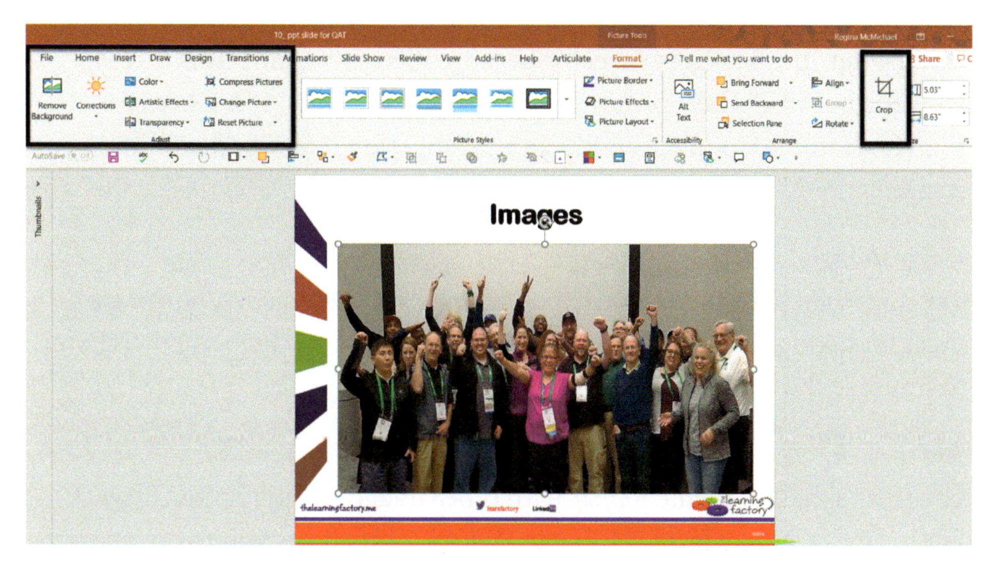

FIGURE 10.10 **Editing images on your Format ribbon**

insert a picture, click on it, and roll over the options on the Format ribbon. It won't take long to master these skills because PowerPoint has made it so easy.

Crop

Use this tool to focus your image on exactly what you want to show your learners. If the image is of a person with their hands near a pinch point, crop the image to show only their hands near the pinch point. Remove all the extra stuff—like the person's face—and focus on what you need to teach. You can verbally explain what machine it is as you discuss the hazard. Be advised that the cropping function is on the far right of the ribbon, away from the creative tools.

Artistic Effects

Use these sparingly and to make a point. These options are interesting but unnecessary for most of your training. Sometimes I might use an effect to blur or pixilate an image as a background with bullet points on top of the image.

Color

If you want to make the image black and white or make the image match the company colors, this is the function for you. Again, I use this for visual accents, not for spicing up a boring presentation.

Remove Background

This is a good tool when you want to focus on a part of the image or to "knock out" the background color because it is distracting. This tool is fantastic but works best on simple images and graphics. If it takes more than five minutes to make an image amazing, skip it and find another image.

Corrections

Try this when your pictures are too dark or light. Our smartphones are great for taking pictures, but sometimes we can use the help of PowerPoint to improve the image.

Right-click

Use this to make a menu appear with other image-editing options (see figure 10.11). My favorite is Change Picture. This allows you to replace one

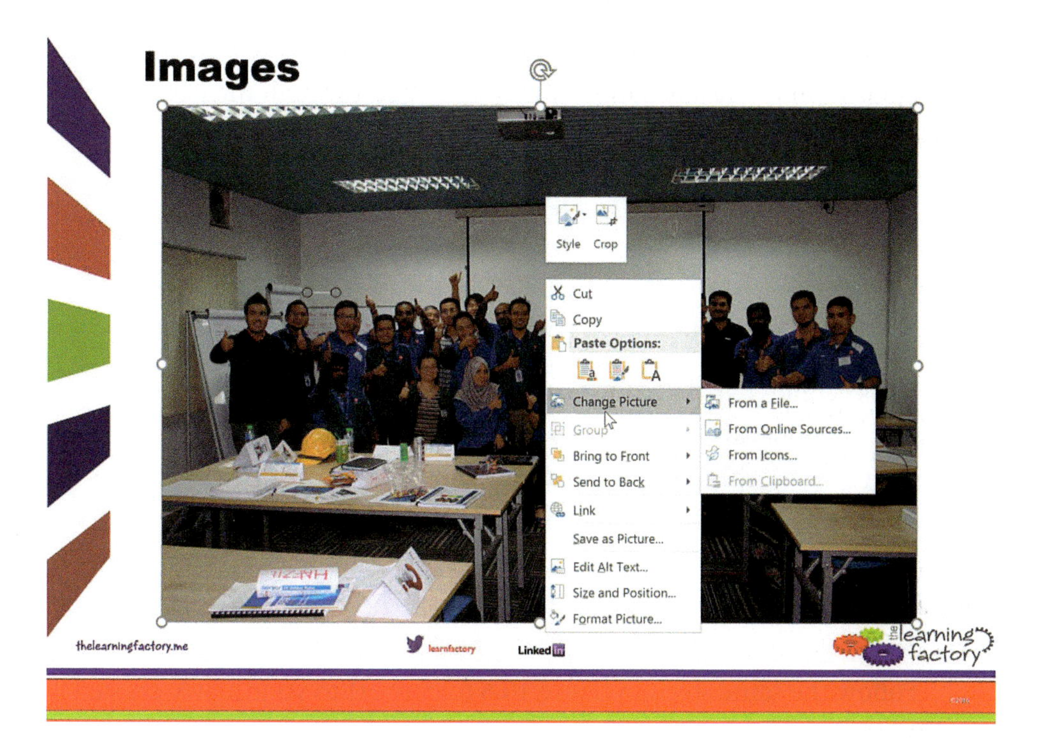

FIGURE 10.11 **Right-click images to edit more**

image with another without losing the spacing or location on the page. If you delete a picture and insert another, you will have to resize it again.

Format Painter and Animation Painter

Format Painter and Animation Painter are two magical tools that can save you time. I have both on my QAT. Format Painter can be found on the Home ribbon, and Animation Painter can be found in the Animation ribbon. Format painter allows you to click on or highlight the words with the format you want, then click on the Format Painter icon, and then run your cursor over the words you want to reformat. This is useful for editing previously created programs if there are a lot of text boxes you are not ready to convert to your new template.

Animation is great when you need to animate actions on screen. I don't think you should animate unless it really adds value to the learning experience. Certainly, don't animate words or sentences so they fly in one at a time—please, I beg you, don't do that! I once used animation to show the sequence of an accident over several slides. The incident involved dropping a tool from over 100 feet (30.5 meters). I cropped the image of the tool and showed it "falling" via animation from the height, through the newly created hole in the roof of the building below, to where it landed right near a work station. It was pretty crude but clearly emphasized the path of the tool from the perspective of the height and the work station. This helped the learners see the message I was teaching.

Revising Older Safety Training Programs

Sadly, this is often harder than starting from scratch. Old artifacts, text boxes, animations, and more will not easily convert into your newly designed templates. This may be good news in disguise because it will give you a chance to ask yourself if this is really what the training should say. For a quick test to determine how much work the conversion will require, simply convert to a new template and see how much changed to the new design.

Follow the same steps in figure 10.6 to convert to your newly created template. If a lot of the information still looks unformatted, it could be because of all those artifacts and text boxes. If you have mastered the Format Painter tool, you will be able to fix fonts a little faster. Don't leave old info in text boxes; take the time to cut and paste it into the actual template boxes already there. It's more work up front, but it will be faster to edit all future versions of the program. I recommend creating a new PowerPoint in your properly developed template and then cutting and pasting only the information you really need from an older, poorly formatted presentation. If you can't do that, be ready to clean up all the slides you converted.

One of the other ways to clean up existing PowerPoints is by making sure each slide is converted to the correct slide inside your template. If you cut and paste fifty slides into your new template, they will likely default to one slide design type. They were probably created that way, so that's how they will arrive in your template. Because you have your new template, you may need to reassign the slide types you now want. As mentioned earlier, my slide types have different names and styles. Figure 10.12 shows an easy way to reassign the slide design you want for each slide. You can also highlight all the slides in the slide view and change all the designs at once.

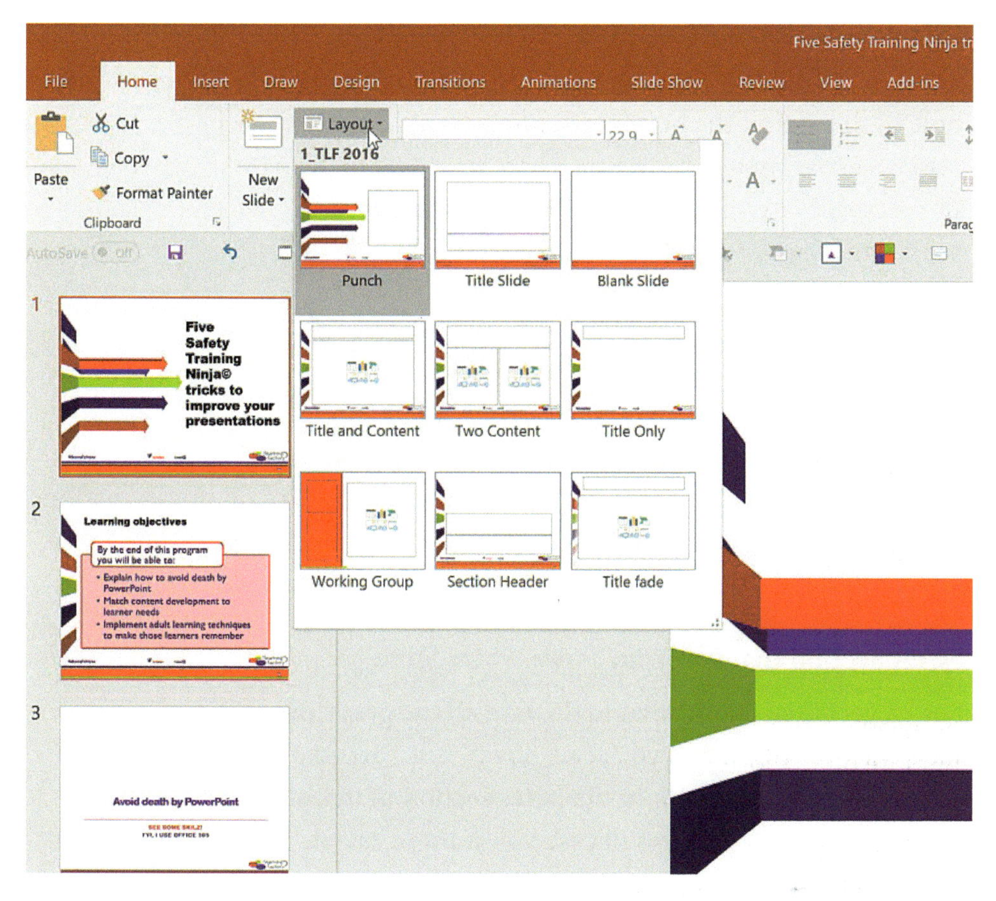

FIGURE 10.12 Changing slide types on the Home ribbon

To the left of the layout icon is the New Slide icon. This is how you insert new slides into your program. If you click only the icon, it will insert the same type of slide you last inserted. If you want to insert another style, click the dropdown button for all your template choices. It's easy to do since this is all on your Home ribbon.

If your program came with a transition you want to remove from every slide, just head to the Transition ribbon, highlight every slide in the slide view, click the None icon, and then click Remove All on the far right (see figure 10.13).

Do you need to remove those horrible animations someone thought were a great idea in the 1990s? You can't remove them all as easily as you can remove transitions. You have to go to each slide, click on the Animation ribbon, highlight the entire slide, and then click None. Don't have time to do that to

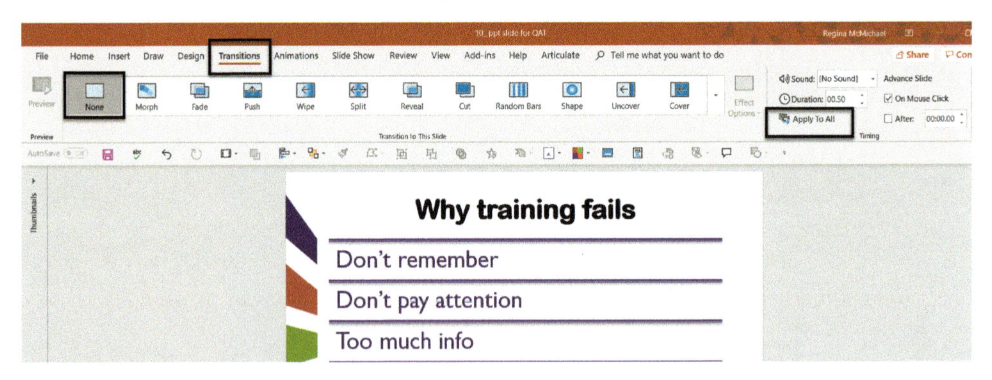

FIGURE 10.13 **Editing transitions on the Transition ribbon**

a huge program? Here is an awesome trick I use. Just turn off the animations! Yes, you can do that! Go to the Slide Show ribbon, click Set Up Slide Show, then under Show Options, check the Show Without Animation box.

Notes Section Beneath Your PowerPoint Slide

I often get asked about what to do with all the great information on the slides that I recommend you scale down. Cut paragraphs and overly worded sentences and paste them in the notes section of that slide. Now you have all the backup information in case you want to review it before the presentation, but it's not on the screen to baffle, bore, or discourage the learners. I also recommend putting your material references, such as the rights to use images, books, articles, or URLs, in the notes section as well. This will help you later for review, and it serves as a source for permissions for anyone who may want to know about them. This extra information is also helpful if you or someone else is auditing your training. Now they will know your sources and what you covered in the training via your bullet points.

The opportunities are endless with PowerPoint, but so is the damage you can cause. So take all this new information and combine it with the other caveats about making sure what you do is for the learners and bridging the learning gap. Writing this chapter made me a better ninja. Some of the things I wrote about here I learned while doing research for this book. As software updates keep on coming, there will always be something new to learn. Keep visiting and subscribe to my YouTube channel for ninja tips and tricks.

Conclusion

Using ADDIE is a big step toward improving the professional content of our EHS training. It's a process, and it will take some work, but it is worth it for your learners and for you. The ADDIE cycle was once considered rigid, but with the experience an EHS professional has with the concepts of continuous improvement, it can become a flowing cycle that will constantly improve your training efforts.

ADDIE Elements

Let's recap the major points of ADDIE and how you can continue to leverage the benefits.

Analyze

Determine what your learners need to change or improve from your training. Find what the learning or knowledge gap is, and limit your training program to just that gap. Gather all the information you can about what they need to know using all the tools available.

Design

Create learning objectives that will fill the identified learning gap. Consider how you will use adult learning principles to make the materials memorable and to stick in the learners' brains after the training is complete. Open your mind to different training offerings beyond the typical lecture, and create your road map for your training program.

Develop

Gather all the content-rich materials you can from sources inside and outside of your organization. Think about how you can avoid the classic compliance-based training program with a lot of words on the PowerPoint screen. Use activities and learner-inclusive techniques that will promote better learning and better retention of the learning. Create job aids and other materials outside of PowerPoint that will help learners apply what they learned after the training program is completed.

Implement

Be the best ninja trainer you can be. Take some risks with your own style and delivery techniques and get the learners as excited as you are about great EHS training. Practice your new skills and prepare yourself properly with the checklist and tools provided in this book.

Evaluate

Beyond the basic smile sheet, consider all the ways you can determine if your training program was effective. After the training, develop tools and job aids and follow up with supervisors and learners whenever possible. Be ready to use what you have learned to continually improve workers' behaviors and ensure their safe work after the class is completed. Take credit for improved safety from your training by keeping records of how you developed and delivered the materials. Use the data and information from the ADDIE process to influence stakeholders and receive more time and/or money for future program development.

Making Immediate Changes

Early in the book, I shared with you the most important ninja trick I want you to remember: When it comes to training design, delivery, and implementation, it's about the learner—always about the learner and not about the trainer.

Throughout the book I asked you to question yourself on your skills, techniques, and even your attitude about your safety training programs. Many of you are ready to change, and that's why you read this book. I applaud you for your great attitude and your desire to be a better trainer for the benefit of those you are trying to protect.

There is a lot of information to process, and it is unlikely that anyone has the time and money to utilize all the phases of ADDIE at once. But you can begin to make some changes right away. Take the time to find some quick, little wins in those changes you can implement. Consider the following things that you could do right now:

- Make sure you are training on only the materials and content your learners need to know.
- Practice writing and using learning objectives in a smaller training program under development.
- Design an icebreaker that shows the value of the training program and energizes your learners at the beginning of the class.
- Replace one lecture section with an activity. For example, demonstrate how to clean and store a respirator instead of lecturing on the regulation's requirements for cleaning and storage.
- Review your PowerPoints and take out anything your learners don't need to know.
- Review your PowerPoints and replace all those long sentences with short bullet points.
- Advertise you and your program as new or different.
- Make a short video with your phone that shows the safe way to do the task you are training on and show it during training.
- Remove all useless clip art, pictures, and copyrighted materials you did not get permission to use.
- Connect with learners and ask if they will help you create a better training program.
- Connect with others in your organization who can help you make a better training program.
- Go online and watch videos on how to be a better trainer or how to use PowerPoint more effectively.
- Follow up after the class to determine if the training sticks with the learners.
- Use one of the checklists in the book.
- Replace a test with another learning confirmation tool, such as a crossword puzzle.

These are just a few simple things you can do right away. After that, develop a game plan to further pursue the use of ADDIE in your organization. Putting the learners' needs first while working toward better EHS training will make our profession and the hard work we do every day even more valuable.

Our Exponential Impact

I will never stop learning or being amazed at the character and energy of safety trainers. As I was in final editing of *The Safety Training Ninja*, I was awestruck by the eight-hour ninja class I facilitated at the National Safety Congress in Houston, Texas. The learners didn't just sit and wait for this ninja to teach—oh no, they immediately dove right in. Within minutes we were behind schedule because they had so much to share and to learn from each other. I stood back and watched the magic. Yes, these were trainers wanting to be better trainers, so it was the best possible scenario, but this group was special.

As they introduced themselves, I was struck by how many people they must touch with their work in safety and teaching. On a whim I asked them to write down on a flip chart how many people each one of them impacted via their safety training. By the end of the first break the number was on the wall, and I was brought to tears. The twenty-five learners in that room could reach and impact 1.4 million people. As I expressed my honor to teach them, with a cracking voice and tears, someone shouted that it was more than that because they impacted their families too! At that point, I was not the only one in tears. Both men and women were profoundly struck by what great training can really do for our workers, contractors, volunteers, the public, our families, and, well, everyone! I honored that awesome class of ninjas by using our class picture in figure 10.10.

That day I became even more positive that training other safety trainers was exactly what I was supposed to be doing in my career. I hope that despite budget constraints, grouchy learners, impatient bosses, uncooperative SMEs, and software limitations, you will do everything you can to be an outstanding safety training ninja because people are counting on you, even if you never personally meet them.

Go do great things!

Further Reading

American Society of Training and Development. *The Transfer of Learning Skills*. Alexandria, VA: American Society of Training and Development, 1998.

ANSI/ASSP. *Criteria for Accepted Practices in Safety, Health and Environmental Training*. ANSI/ASSP Z490.1-2016. Park Ridge, IL: American Society of Safety Professionals, 2016.

Bean, Cammy. *The Accidental Instructional Designer*. Alexandria, VA: American Society for Training & Development, 2014.

Biech, Elaine. *The Art and Science of Training*. Alexandria, VA: Association for Training Development, 2017.

Biech, Elaine. *Training and Development for Dummies*. Hoboken, NJ: Wiley, 2015.

Bloom, Benjamin S. *Taxonomy of Educational Objectives, Book 1: Cognitive Domain*. New York: David McKay, 1956.

Board of Certified Safety Professionals. "Certified Environmental, Safety and Health Trainer." Certified Environmental, Safety and Health Trainer (CET). Accessed October 18, 2018. https://www.bcsp.org/CET.

Carliner, Saul. *Designing E-learning*. Alexandria, VA: American Society for Training & Development, 2006.

Carliner, Saul. *Training Design Basics*, Second Edition. Alexandria, VA: Association for Talent Development, 2015.

Chapman, Bryan. "How Long Does It Take to Create Learning?" Chapman Alliance, September 2010. http://www.chapmanalliance.com/howlong/.

Cohen, Alexander, and Michael J. Colligan. *Assessing Occupational Safety and Health Training: A Literature Review*. DHHS (NIOSH) Publication No. 98-145. Cincinnati, OH: NIOSH, June 1998. https://www.cdc.gov/niosh/docs/98-145/pdfs/98-145.pdf.

Escribano, J. C. "Learning Circuits Big Question: Working with Subject Matter Experts." LifeLongLearningLab. Accessed August 31, 2018. https://mylifeismylab.wordpress.com/2009/09/28/learningcircuits-big-question-working-with-subject-matter-experts/.

Gutierrez, Karla. "Attention-Grabbing eLearning Design: 5 Techniques You Should Try." SHIFT ELearning Blog, May 17, 2016. https://www.shiftelearning.com/blog/attention-grabbing-elearning-design.

Kaveti, Vandana. "Why Use Job Aids in Training?" CommLab India, April 11, 2012. https://blog.commlabindia.com/elearning-design/job-aids-in-training

Kirkpatrick, James D., and Wendy Kayser Kirkpatrick. *Kirkpatrick's Four Levels of Training Evaluation*. Alexandria, VA: Association for Training Development, 2016.

Krebs, Erin. "Job Aids . . . and If You Need Them." SweetRush. December 13, 2012. https://www.sweetrush.com/job-aids-and-if-you-need-them/.

Moore, Cathy. "How to Convert the Toughest SME," March 30, 2010. http://blog.cathy-moore.com.

Pappas, Christopher. "7 Tips to Beat Short Attention Spans In eLearning." eLearning Industry, May 23, 2015. https://elearningindustry.com/7-tips-to-beat-short-attention-spans-in-elearning.

Russo, Cat, and Jennifer Mitchell. *Dictionary of Basic Trainer Terms*. Alexandria, VA: American Society for Training & Development, 2006.

Snyder, Daniel J. *CET Exam Study Workbook*. Nixa, MO: SPN International Training, 2018.

Sullivan, Richard L. *The Transfer of Skills Training*. Alexandria, VA: American Society for Training & Development, 1998.

Thalheimer, Will. *Performance-Focused Smile Sheets: A Radical Rethinking of a Dangerous Art Form*. Somerville, MA: Work-Learning Press, 2016.

Thomas, Willis H. *Templates for Managing Training Projects*. Alexandria, VA: American Society for Training & Development, 2015.

Index

I

J

K

L

M

N

About the Author

Regina McMichael, CSP, CET, is President of The Learning Factory Inc. She has twenty-nine years of experience in safety and health education, training, communication, and leadership. Regina inspires organizations around the world with her messages of combining compassion and the business of safety. She also mentors safety professionals to achieve their dreams of becoming Safety Training Ninjas and passion-driven professionals.

Regina is also the founder of Peace, Love & Safety, the leadership and training catalyst supporting organizations while elevating their EHS programs from compliance-driven checklists to human- and humanity-based solutions. Regina is an active volunteer in the safety and health profession, serving on the Board of Certified Safety Professionals, the American Society of Safety Professionals, and the National Safety Council. She is also an active member of the Association for Talent Development and the National Speakers Association. Her diverse background has allowed her to provide hands-on expertise in the research, writing, and production of books, training manuals, videos, and articles.

Regina has worked in the construction, insurance, aerospace, automotive and service industries. She loves traveling the world to deliver her safety and leadership messages to all types of professionals and workers. You can reach her at regina.mcmichael@gmail.com.